W0260463

ISBN 978-3-540-04115-3 ISBN 978-3-540-35815-2 (eBook)
DOI 10.1007/978-3-540-35815-2

**Fortschritte der chemischen Forschung** 10. Band, 3. Heft



---

This series presents critical reviews of the present position and future trends in modern chemical research. It is addressed to all research and industrial chemists who wish to keep abreast of advances in their subject.

As a rule, contributions are specially commissioned. The editors and publishers will, however, always be pleased to receive suggestions and supplementary information. Papers are accepted for „Fortschritte der chemischen Forschung" in either German or English.

Single issues may be purchased separately.

In kritischen Übersichten werden in dieser Reihe Stand und Entwicklung aktueller chemischer Forschungsgebiete beschrieben. Sie wendet sich an alle Chemiker in Forschung und Industrie, die am Fortschritt ihrer Wissenschaft teilhaben wollen.
In der Regel werden nur Beiträge veröffentlicht, die ausdrücklich angefordert worden sind. Schriftleitung und Herausgeber sind aber für ergänzende Anregungen und Hinweise jederzeit dankbar. Manuskripte können in den „Fortschritten der chemischen Forschung" in Deutsch oder Englisch veröffentlicht werden.

Jedes Heft der Reihe ist auch einzeln käuflich.

---



---

# The Chemistry of Biguanides

**Dr. F. Kurzer and Dr. E. D. Pitchfork**

Royal Free Hospital School of Medicine (University of London), 8 Hunter Street, London, W.C. 1, England

**Table of Contents**

# I. Introduction

The chemistry of biguanides has provided a fertile field of chemical research ever since the first preparation of the parent base by *Rathke* (*524*) in 1879. However, this interesting branch of organic chemistry has apparently so far not been the subject of a comprehensive review. The present summary is intended to survey very briefly all aspects of the existing knowledge of biguanide chemistry. Particular attention is given to the striking advances of the last 25 years, but earlier work is referred to whenever it provides the necessary background to more recent progress.

The study of the chemistry of biguanides received renewed impetus by the realisation that 1-p-chlorophenyl-5-isopropylbiguanide ("Paludrine") is an effective antimalarial drug. Outstanding contributions were made by *Curd*, *Rose*, and their co-workers, who reported, between 1946 and 1950, the results of their reinvestigation and extension of many aspects of biguanide chemistry in a memorable series of papers (*546*). During the same decade, much interest was shown by the American Cyanamid Company in the synthesis of guanamines by the interaction of biguanides with esters. More recently, biguanides have once again attracted widespread interest as oral hypoglycemic drugs. However, by the time the most useful of these, 1-β-phenetylbiguanide, had been firmly established (*726*) in 1957, little chemical work remained to be done, and the relevant information has therefore appeared mainly in the pharmacological literature.

A discussion of metal complexes of biguanides has recently formed part of an extensive review (*532*) on co-ordination compounds, so that this subject need not be considered here in detail. As in the past, Indian laboratories continue to be the main source of publications on this aspect of biguanide chemistry (*38, 193, 195, 281, 282, 300, 435, 527, 528, 529, 533, 586, 649, 650*).

In view of the manifold interest and obvious medical value of biguanides it is not surprising that a large proportion of the available information is embodied in patent specifications, many of them of fundamental scientific interest. Although investigations on biguanides have been diverse, much remains to be done. It is hoped that the present account will go some way towards stimulating further interest in the chemistry of these reactive and versatile compounds, and that some of the inconsistencies and omissions in the literature will become the subject of further studies.

### *Nomenclature*

In conformity with the practice of *Chemical Abstracts* and most major journals, the name *biguanide* is adopted in the present review. The term *diguanide* is also used, particularly in British publications.

The following accepted mode of numbering the molecule enables all derivatives to be named unequivocally:

$$\overset{1}{H_2N}\cdot \overset{2}{C}(:NH)\cdot \overset{3}{NH}\cdot \overset{4}{C}(:\overset{5}{NH})NH_2$$

The name *biguanidine* occasionally appears in the literature in error; this should of course apply to the distinct class of compounds based on the parent structure $NH_2C(:NH)\cdot NHNH\cdot C(:NH)NH_2$.

### *Structure*

The structure of biguanide (I), whether regarded as a condensed system of amidino-groups or of guanidine-units, reflects the chemical versatility and high reactivity of this series of compounds. The close relationship of biguanides with amidino(thio)ureas, cyanoguanidines, biurets (and hence indirectly with guanidine and urea) is demonstrated by the existence of comparable syntheses, and by the general physical and chemical behaviour of relevant compounds. Because of its particular chain-length, the biguanide-grouping is able to participate in a variety of simple cyclisation reactions.

In common with other compounds related to urea, biguanide cannot be adequately defined by a single structural formula, but is represented as a resonance hybrid of contributing mesomeric forms. Because of the remarkable accumulation, in the biguanide pattern, of groupings capable of undergoing prototropy and mesomerism, the number of theoretically possible canonical structures is large. Thus, the simplest biguanide structure (I) may resonate among seven mesomeric forms (I a—g). The molecule may assume further tautomeric configurations (e.g. I h), including those of zwitter-ions (e.g. I i, j), and each of these forms may in turn undergo mesomerism. Moreover, the structure may be stabilised as six-membered configurations, such as (I k), by hydrogen bonding.

A knowledge of the site of protonation in biguanides is of particular interest, because of the importance of such cations as reactive entities, and their possible function in physiologically active biguanides. The problem has attracted a great deal of attention (see Section VI) and has been the subject of critical discussion (*206, 227, 602*). A recent study

(*742a*) of the n.m.r. spectra of several biguanide derivatives has provided evidence in favour of the highly symmetrical structures I l and I m for the mono- and di-cations of biguanide. The tri-cation has several possible prototropic structures, one of which is shown (I n). It is recalled that guanidine is a particularly strong base since its cation,

$$\delta+H_2N\cdots C\begin{matrix} \cdots NH_2\ \delta+ \\ \cdots NH_2\ \delta+ \end{matrix}$$

having its charge spread uniformly by mesomerism, is perfectly symmetrical and has a very large resonance energy. The high basic strength of biguanide may be ascribed to a similar origin.

The numerous highly coloured metal complexes obtainable from biguanides and elements of the transition series have posed yet another interesting structural problem. The accumulated evidence suggests that the structures agreeing most satisfactorily with the observations are of type (I o; M = divalent metal). This aspect of the subject has been discussed fully by *Ray* (*532*).

A recent X-ray-crystallographic investigation (*86*) has provided the first accurate structural picture of a biguanide derivative, namely that of the antimalarial Paludrine (1-p-chlorophenyl-5-isopropylbiguanide hydrochloride) (Fig. 1). The biguanide part of the cation consists of two

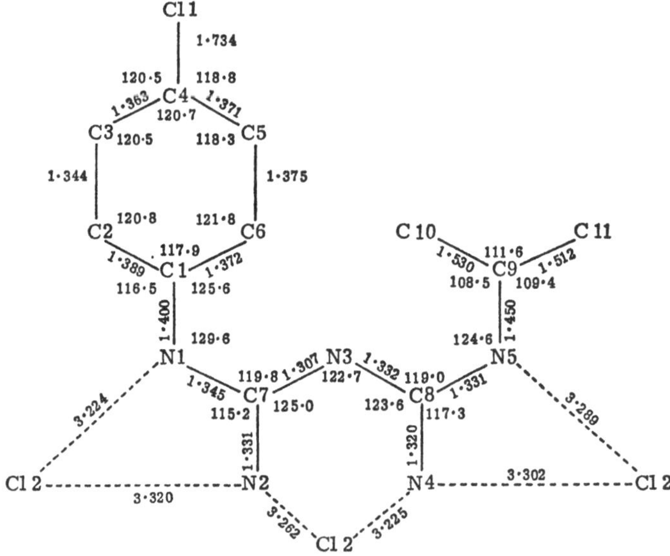


Fig. 1. Bond lengths and interbond angles of Paludrine (*86*). (*Reproduced by kind permission of the author, and of the Chemical Society, London*)

planes of atoms, each containing a $CN_3$-group, intersecting at an angle of 58.9°. The mean C—N-distance being 1.33 Å, there is no distinction between formal single and double bonds, thus providing evidence for the mesomeric nature of the structure.

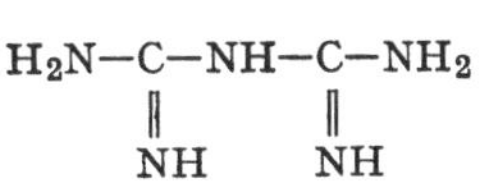

I   Ia   Ib

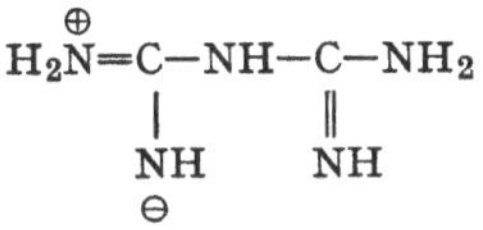

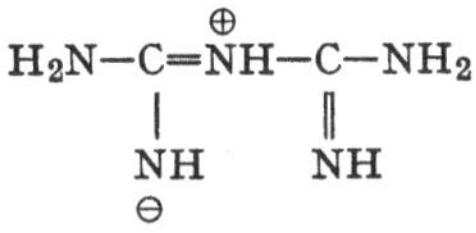

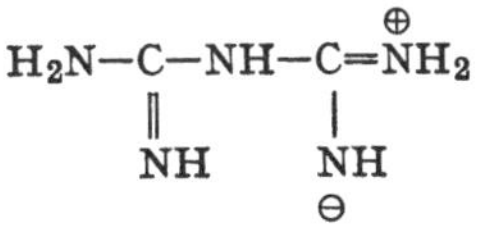

Ic   Id   Ie

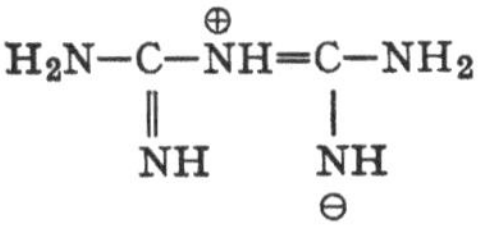

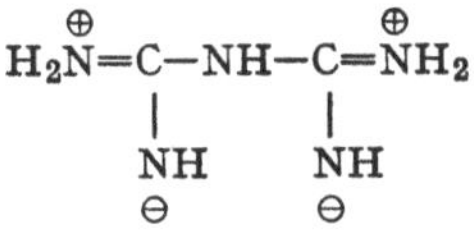

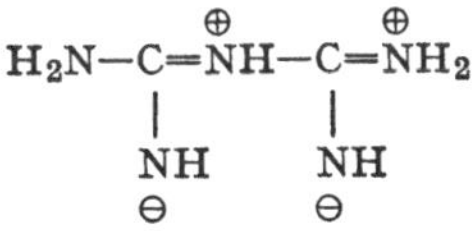

If   Ig   Ih

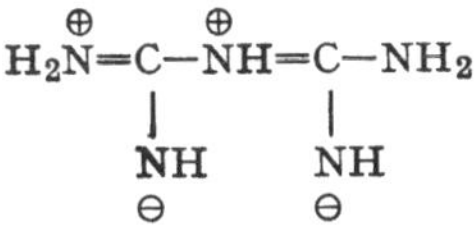

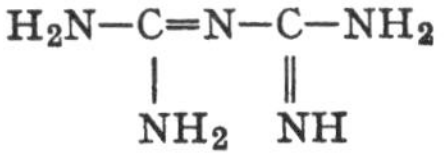

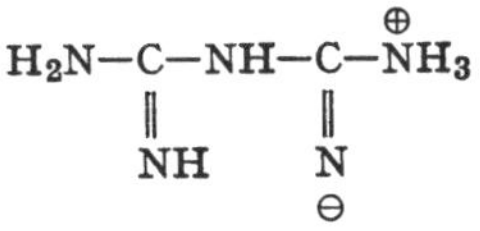

Ii   Ij   Ik

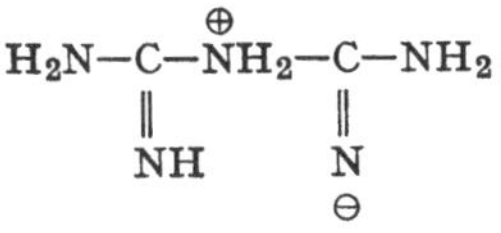

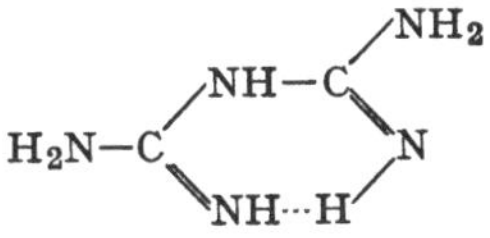

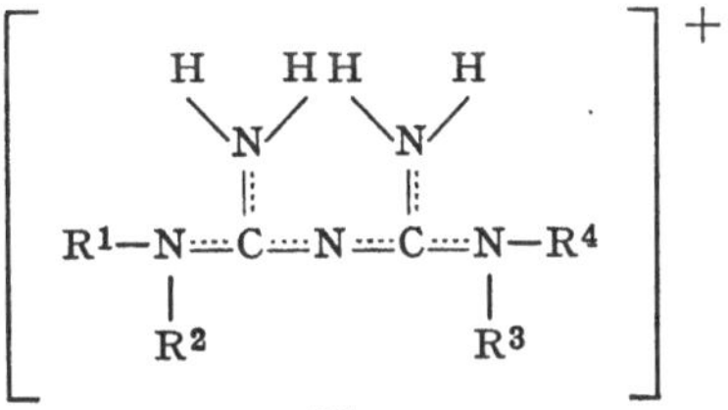

Il   Im

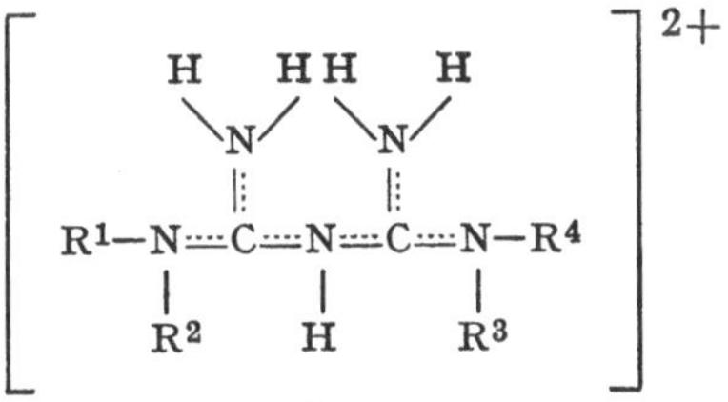

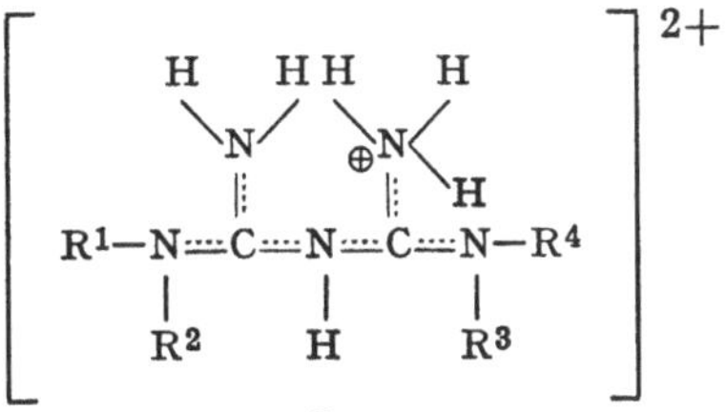

In   Io

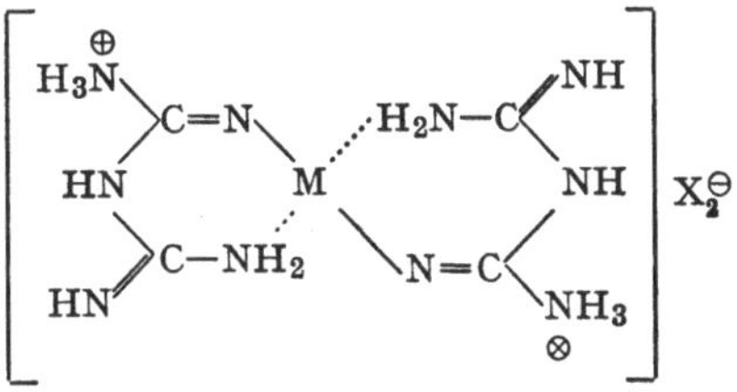

## II. Synthesis of Biguanide

Existing syntheses of the parent compound are listed in Beilstein's Handbook.

A recent new synthesis of biguanide, affording 50—60% yields of product, utilises the reaction between O-alkylisoureas and equimolar proportions of guanidine and its hydrochloride in ethanol (*612*). Small amounts of melamine and cyanoguanidine are formed as by-products.

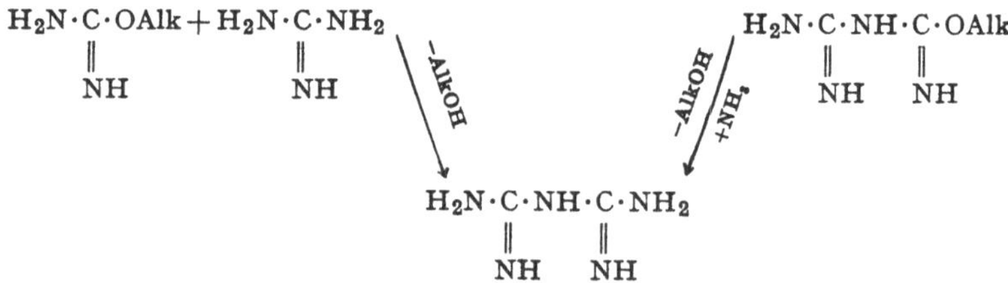


A related approach to biguanide is the ammonolysis of O-methyl-amidinoisourea (*352, 354*). The hydrochloride, when kept in 28% ammonium hydroxide solution at 15—20° for 20 hours, is reported to produce biguanide in 98% yield.

Detailed descriptions of the preparation of biguanide sulphate (*347*) and of the isolation of the free base therefrom have recently appeared in *Inorganic Syntheses* (*325, 436*).

*Cyanobiguanide*

With acetone and hydrogen chloride, cyanoguanidine forms the addition compound II; this yields melamine on treatment with alkalis at room temperature, cyanamide being eliminated. In the presence of copper or nickel hydroxide, an intermediate cyanobiguanide-complex salt can be isolated, and converted by dilute nitric acid at 20° into cyanobiguanide

$$(NH_2\cdot\underset{\underset{NH}{\|}}{C}\cdot NHCN)_2\cdot Me_2CO\cdot 2HCl \xrightarrow{OH^\ominus} NH_2\underset{\underset{NH}{\|}}{C}NH\underset{\underset{NH}{\|}}{C}\cdot NHCN$$

(II) (III)

$$\left[NH_2-\underset{\underset{\oplus NH_2}{\|}}{C}-NH-\underset{\underset{NH_2}{|}}{C}{=}NCN\right] NO_3^\ominus \qquad NH_2-\underset{\underset{NH_2}{|}}{C}{=}N-\underset{\underset{NH_2}{|}}{C}{=}NCN$$

(III a) (III b)

nitrate (III) (*51*) (see also ref. *758*). Infra-red (*51*, *453*) and ultraviolet (*453*) spectra suggest that the structures of the nitrate and free base in media of pH 6—9 are IIIa and IIIb respectively.

## III. Synthesis of Substituted Biguanides

### A. Ammonolysis of Cyanoguanidines

The fusion of cyanoguanidine with ammonium salts is one of the earliest and most widely used syntheses of the parent compound, biguanide (*31, 419, 482, 630, 631*). The nature of the ammonium salt used influences the yield [e.g. $NH_4Cl$, 21—23% (*31*); $NH_4I$, 45% (*482*)] but the view that this effect is attributable to the lowering of the fusion temperature (*203, 668*) has not been confirmed (*488*). *Rackmann's* (*518*) original ammonolysis of cyanoguanidine, using aqueous ammonia under pressure at 110° in the presence of cupric sulphate, has recently been reported to yield amidinourea rather than biguanide. Attempts to apply *Rackmann's* procedure to N-arylsulphonyl-N′-cyanoguanidines gave only 7—12% of the required biguanides, the majority (30—40%) of the starting material remaining unchanged, while part (12—32%) was hydrolysed to the sulphonylguanidines (*379*).

More recently, ammonium sulphonates have proved excellent reagents for converting cyano- into amidino-groups (*488*). Applied to the present case, the use of ammonium sulphonates at 140—160° is particularly useful in the production of 1-arylbiguanides (*44, 488, 613, 614*). As expected, action of two moles of ammonium sulphonates results in guanidines.

(IV) [cyclic structure]

$CH_2 \cdot NH \cdot C(=NH) \cdot NH \cdot C(=NH) \cdot NH_2$ / $CH_2 \cdot NH \cdot C(=NH) \cdot NH \cdot C(=NH) \cdot NH_2$ (V)

$CH_2 \cdot NH \cdot C(=NH) \cdot NH \cdot C(=NH) \cdot NH_2$ / $CH_2OH$ (VII) $\xrightarrow{-H_2O}$

$CH_2NH_2$ / $CH_2NH_2$ $\longrightarrow$ $CH_2 \cdot NH \cdot C(=NH) \cdot NH \cdot C(=NH) \cdot NH_2$ / $CH_2NH_2$ (VI) $\xrightarrow{-NH_3}$ (VIII) [imidazoline]$NH \cdot C(=NH) \cdot NH_2$

The fusion of equimolar amounts of ethylene diamine hydrochloride and cyanoguanidine gives, according to conditions, different products that have been the subject of some controversy (*99, 184, 190, 530*). *Dittler's* (*184*) original results and formulation (IV) have not been substantiated. The reaction yields in fact (*99, 190, 510*) 2-guanidinoimidazole (VIII), also obtainable by dehydration using fuming sulphuric acid of β-hydroxyethylbiguanide (VII) (*510*), or by heating ethylenedibiguanide (V) (*510*). In presence of cupric sulphate and *two* moles of cyanoguanidine, the dibiguanide (V) formed is stabilised by complex-formation and becomes the main product of the reaction (*99*).

*Biguanides as Reaction Intermediates*

The intermediate existence of biguanide in the synthesis of guanidine salts by fusing cyanoguanidine and ammonium salts has frequently been discussed (*68, 173, 419, 629*), and has recently been reconsidered (*489, 668*). It is recalled, in support of this view, that cyanoguanidine and aqueous ammonia react in the presence of cupric sulphate to form a copper-biguanide complex (*518*). Under these conditions, the copper presumably removes the biguanide as soon as it is formed, and prevents further breakdown. It is thus widely accepted that biguanide salt is primarily formed, and decomposes under the influence of a second molecule of ammonium salt to yield two moles of guanidine salt (Path a). *Sugino* (*668*), however, has provided some observations suggesting that biguanide salt decomposes by itself either to a guanidine salt (Path b) or back to cyanoguanidine.

$$H_2N\cdot\underset{\underset{NH}{\|}}{C}\cdot NHCN + NH_3HX \longrightarrow \left[ H_2N\cdot\underset{\underset{NH}{\|}}{C}\cdot NH\cdot\underset{\underset{NH}{\|}}{C}\cdot NH_2\cdot HX \right] \xrightarrow[\text{(a)}]{NH_3HX} 2\,H_2N\cdot\underset{\underset{NH}{\|}}{C}\cdot NH_2\cdot HX$$

$$\downarrow \text{(b)}$$

$$H_2N\cdot\underset{\underset{NH}{\|}}{C}\cdot NH_2\cdot HX + NH_2CN$$

Later workers (*489*) have postulated the existence of intermediate biguanide in the interaction of cyanoguanidine and ammonium nitrate in liquid ammonia (at relatively high temperatures). The reaction is thus thought to adopt the same reaction path as that of the analogous fusion, but positive evidence for this view is difficult to obtain, because of the rapidity with which all postulated intermediates are onverted into guanidine salts.

The fusion of cyanoguanidine and ammonium nitrate in the molar proportions 1 : 2·4 has recently been reinvestigated quantitatively (*356*) by tracing the dependence on time of the melt-content of biguanide, guanidine nitrate and cyanoguanidine at three separate temperatures (130°, 140°, and 150°). At each temperature, the formation of biguanide nitrate starts as soon as the reaction mixture is placed in the thermostat. Its content in the melt rises to a maximum and then diminishes, the decrease setting in when its formation rate becomes less than that of guanidine nitrate.

Fusion of 1-methyl- or 1,1-dimethyl-3-cyanoguanidine with ammonium nitrate at 120—150° produces 1-methyl- or 1,1-dimethylbiguanide (*362*) as expected.

## B. Aminolysis of Cyanoguanidines

The action of amines on cyanoguanidine, the logical extension of the ammonolysis described in the foregoing section, is undoubtedly the most widely used route to substituted biguanides. Numerous examples of its broad applicability occur throughout the literature (*9, 27, 32, 35, 36, 41, 52, 305, 313, 338, 426, 427, 473, 534, 547, 558, 562, 581, 671, 685, 693*), many of them in the form of patent specifications (*14, 15, 106, 158, 185, 222, 383, 424, 444, 445, 515, 596—598, 724*).

$$\underset{\underset{\text{NH}}{\|}}{\text{RNH}\cdot\text{C}}\cdot\text{NHCN} + \text{R}'\text{NH}_2 \rightarrow \text{RNH}\cdot\underset{\underset{\text{NH}}{\|}}{\text{C}}\cdot\text{NH}\cdot\underset{\underset{\text{NH}}{\|}}{\text{C}}\cdot\text{NHR}'$$

The reaction may be carried out successfully by the conventional fusion-technique. In the course of the synthesis of numerous aryl- and alkylbiguanides by this procedure, the substituted guanidine rather than the biguanide was occasionally formed in a side reaction. In such cases the biguanide first formed may be cleaved partially by way of an intramolecular hydrogen-bonded form (*602*) (cf. section VI, B3).

$$\overset{\oplus}{\text{RNH}_2}\text{–C}(\text{–NH}_2\cdots\text{H})\text{=N–C}(\text{–NH}_2)\text{=N}\cdots \longrightarrow \left[\overset{\oplus}{\text{RNH}_2}\text{–}\underset{\underset{\oplus\text{NH}_3}{|}}{\text{C}}\text{=}\overset{\ominus}{\text{N}} + \overset{\oplus}{\text{C}}(\text{–NH}_2)\equiv\overset{\ominus}{\text{N}}\right] \longrightarrow \text{HR}\overset{\oplus}{\text{N}}\text{H}\cdot\underset{\underset{\text{NH}}{\|}}{\text{C}}\cdot\text{NH}_2 + \underset{\underset{\text{CN}}{|}}{\text{NH}_2}$$

A variation of the procedure employing an excess of acid obviates the need for fusion of the reactants and allows aqueous media to be used (*735*). This technique has been thoroughly investigated by *Curd*, *Rose* and their co-workers. Interaction of arylamine salts and aromatic cyanoguanidines in boiling aqueous media proceeded in good yields (*142, 149, 150, 361*), the biguanide salt often separating directly on cooling. The optimum amount of water is between 12.5 and 18% of the total weight of the equimolar mixture (*728*). The reaction is accelerated (*25, 148*) in aqueous dioxan or cellosolve, and is catalysed (*149*) by zinc or copper salts.

Tertiary bases (e.g. pyridine or dimethylaniline) are preferable as solvents to water (*335, 336*) for condensing certain aromatic amines and the parent, cyanoguanidine. When the amine concerned lacks acidic substituents, an equivalent of a mineral acid is a useful catalyst (*336*): it is best introduced as a salt of the amine or a tertiary base, or as gaseous hydrogen chloride, to maintain anhydrous conditions. If the amine does contain acidic substituents an acid catalyst is unnecessary (*335*).

Aliphatic amines and arylcyanoguanidines interacted slowly (*148*), even in the higher-boiling solvents, but were condensed successfully in aqueous ethanol containing cupric sulphate or copper bronze (*148, 151*). This method was finally superseded by the introduction of even higher boiling solvents, e.g. nitrobenzene (*143, 232, 456*) when good yields of the desired biguanides were obtained in numerous examples (*456*). Once again, copper or zinc salts are effective as catalysts (*152, 154*).

The condensation of alkylcyanoguanidines and arylamines (*143, 154*) (i.e. alkyl and aryl groups "crossed") is important in that certain alkylcyanoguanidines, in contrast to the aryl-homologues, are readily accessible. Boiling water or aqueous or anhydrous cellosolve are the solvents of choice, but direct fusion of the reactants has also been employed (*133*). The general method has been extended to more highly substituted cyanoguanidines (*58, 133, 534, 598*), and to halogeno-substituted amines (*132*).

Following the discovery of the antimalarial activity (*148*) of biguanides and quinoline derivatives, a variety of substituted quinolyl-biguanides were synthesised (*9, 42, 101, 188, 298, 303, 306, 428, 558, 583, 585*) by these procedures. The production of a series of ω,ω-hexamethylenebiguanides (*724*) illustrates the use of secondary amines in this synthesis.

It is noteworthy that cyanoguanidine and p-aminosalicylic acid react in acidified boiling water with simultaneous decarboxylation (*608*), giving m-hydroxyphenylbiguanide instead of the expected 1-(3-hydroxy-4-carboxyphenyl)biguanide. The decarboxylation under such mild conditions appears to be anomalous, since p-aminobenzoic acid (*100*) and its esters (*97*) do give the expected biguanides.

$$CH(OEt)_2\text{—}CH_2\cdot NHR \longrightarrow CH(OEt)_2\text{—}CH_2\cdot NR\cdot C(:NH)NH\cdot C(=NH)\cdot NH_2$$

$$\downarrow$$

$$\text{(VIII)}\ \text{[imidazole ring, N-R]}\text{—}NH\cdot C(=NH)\cdot NH_2 \longleftarrow \left[ CHO\text{—}CH_2\cdot NR\cdot C(=NH)\cdot NH\cdot C(=NH)\cdot NH_2 \right]$$

The condensation of cyanoguanidine and aminoacetal (or its methyl-homologue) yields the expected diethylacetal of biguanidoacetaldehyde; this is cyclisable to 2-guanidinoimidazole (or its 1-methyl-homologue) (VIII) by means of mineral acids (*576*).

An example of the application of a hydrazide in this general synthesis, amongst others (*91, 391, 392, 416*), is the reaction of isonicotinic acid hydrazide and cyanoguanidine in the presence of acid to yield isonicotinoylaminobiguanide [$C_5H_4CO \cdot NHNH \cdot C(:NH) \cdot NHC(:NH) \cdot NH_2$] (*391, 392, 416*). Long-chain fatty acid hydrazides react similarly (*685*).

1-Acylaminobiguanides are cyclised by alkalis without difficulty, with loss of water, to 5-guanidino-1,2,4-triazoles. Their interaction with formic acid affords the expected guanamines in the majority of examples (*391, 392*) (see also Section VII, I). Contrary to a previous claim (*91*), isonicotinic acid hydrazide dihydrochloride and cyanoguanidine do not give the biguanide under neutral conditions, but form merely a molecular adduct of the reactants (*416*).

Sulphonylbiguanides are available from sulphonylcyanoguanidines by this general reaction as follows (*581*):

$$RSO_2Cl + NH_2 \cdot \underset{\underset{NH}{\|}}{C} \cdot NHCN \rightarrow RSO_2 \cdot NH \cdot \underset{\underset{NH}{\|}}{C} \cdot NHCN \xrightarrow{ArNH_2}$$

$$RSO_2 \cdot NH \cdot \underset{\underset{NH}{\|}}{C} \cdot NH \cdot \underset{\underset{NH}{\|}}{C} \cdot NHAr$$

*Sen* and *Gupta* (*581*) have recently claimed the preparation of carbamoylbiguanide (and its sulphur-analogue) by refluxing cyanoguanidine with urea nitrate (or thiourea) in aqueous solution. However, the remarkable ease with which compounds of this type are cyclised to triazines (see Section VII, K and L) would suggest that this report requires further confirmation, as does the alleged synthesis (*581*) of corresponding aryl-derivatives by procedures (see Scheme) that would clearly be expected to lead to ring-closure (see also Section IV).

$$NH_2 \cdot \underset{\underset{NH}{\|}}{C} \cdot NHCN + NH_2 \cdot CX \cdot NH_2 \rightarrow NH_2 \cdot \underset{\underset{NH}{\|}}{C} \cdot NH \cdot \underset{\underset{NH}{\|}}{C} \cdot NH \cdot \underset{\underset{X}{\|}}{C} \cdot NH_2$$

$$\downarrow ArNH_2$$

$$NH_2 \cdot \underset{\underset{NH}{\|}}{C} \cdot NHCN + ArNH \cdot CX \cdot NH_2 \rightarrow NH_2 \cdot \underset{\underset{NH}{\|}}{C} \cdot NH \cdot \underset{\underset{NH}{\|}}{C} \cdot NH \cdot \underset{\underset{X}{\|}}{C} \cdot NHAr$$

*Production of Dibiguanides*

The use of diamines in this synthesis affords the expected dibiguanides (*549, 562*), or a bis-cyanoguanidine may be condensed with two moles of amine (*345, 358, 549*). The use of *both* bis-cyanoguanidine *and* a diamine, however, may give rise to polymeric biguanides that are difficult to fractionate (*549*).

In the above reactions, aromatic amine salts give better results than the free bases (*14, 15*), which are indeed reported to fail to react with cyanoguanidine (*117*). Aliphatic amines react more smoothly in the presence of some free amine (*14, 15*). As aromatic amine salts dissociate to a much larger extent than do aliphatic amine salts, it is believed (*14, 15*), that a considerable amount of free aromatic amine is present in the reaction mixture. Since *Curd* and *Rose* (*148*) have also shown that use of (100%) free aliphatic amine lowered the yields of biguanides, a mixture of free amine and amine salt would appear to be most favourable for biguanide formation.

## C. Ammonolysis of Amidinothioureas

An obvious approach to biguanides is the ammonolysis or aminolysis of the sulphur-function in amidinothioureas (IX), and their S-alkyl-derivatives (XIV). An early example of this reaction is due to *Bamberger* (*30*).

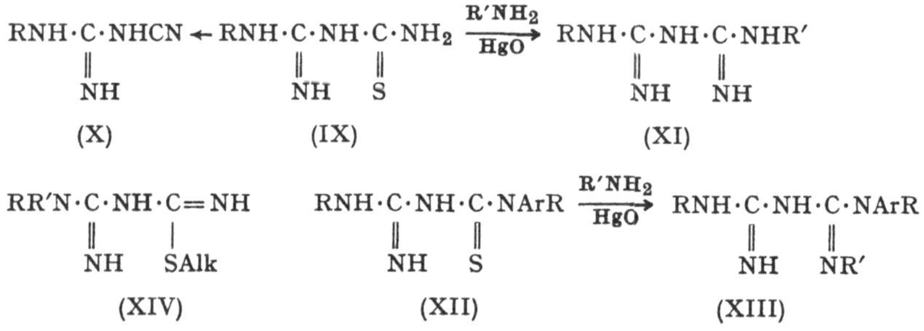

$$\underset{(X)}{\mathrm{RNH{\cdot}\underset{\underset{NH}{\|}}{C}{\cdot}NHCN}} \leftarrow \underset{(IX)}{\mathrm{RNH{\cdot}\underset{\underset{NH}{\|}}{C}{\cdot}NH{\cdot}\underset{\underset{S}{\|}}{C}{\cdot}NH_2}} \xrightarrow[\mathrm{HgO}]{\mathrm{R'NH_2}} \underset{(XI)}{\mathrm{RNH{\cdot}\underset{\underset{NH}{\|}}{C}{\cdot}NH{\cdot}\underset{\underset{NH}{\|}}{C}{\cdot}NHR'}}$$

$$\underset{(XIV)}{\mathrm{RR'N{\cdot}\underset{\underset{NH}{\|}}{C}{\cdot}NH{\cdot}\underset{\underset{SAlk}{|}}{C}{=}NH}} \qquad \underset{(XII)}{\mathrm{RNH{\cdot}\underset{\underset{NH}{\|}}{C}{\cdot}NH{\cdot}\underset{\underset{S}{\|}}{C}{\cdot}NArR}} \xrightarrow[\mathrm{HgO}]{\mathrm{R'NH_2}} \underset{(XIII)}{\mathrm{RNH{\cdot}\underset{\underset{NH}{\|}}{C}{\cdot}NH{\cdot}\underset{\underset{NR'}{\|}}{C}{\cdot}NArR}}$$

### 1. Use of Desulphurising Agents

In an attempt to employ this route *Curd, Rose* et al. (*58*) treated p-chlorophenylamidinothiourea with isopropylamine and mercuric oxide, but obtained mostly N-p-chlorophenyl-N'-cyanoguanidine. Efforts to improve the results by numerous changes in conditions were unsuccessful; in all cases biguanides were merely by-products, formed in yields less than 5%. A later patent (*154*) has however claimed the successful use of this route.

The authors (*58*) consider that the small amounts of biguanides arise from the intermediate cyanoguanidines by further action of amines. Thus, small yields of Paludrine are obtainable from N-p-chlorophenyl-N'-cyanoguanidine and methanolic isopropylamine under similar conditions, either with or without mercuric oxide. Secondly, amidinothioureas give cyanoguanidines on being boiled with mercuric oxide in methanol in the absence of amines, although their presence facilitates the reaction. Finally, amidinothioureas and amines, when reacting under the preferred conditions of this condensation involving the corresponding cyanoguanidines (*143*), gave improved yields of biguanides.

In this synthesis, more fully substituted amidinothioureas of type (XII) are of special interest in that they cannot preferentially form cyanoguanidines by loss of hydrogen sulphide. Thus, the interaction of $\alpha,\omega$-disubstituted amidinothioureas and amines (*133, 154*) in the presence of a desulphurising agent is a flexible general method for the synthesis of biguanides, and has been extended to 1,2,4,5-tetrasubstituted biguanides (*24*). Ammonia similarly produces 1,5-disubstituted biguanides (*154*).

### 2. Use of S-Alkylamidino*iso*thioureas

It is well known that alkylthiol-groups undergo aminolysis and allied reactions more readily than do free mercapto-groups, and S-alkylamidino-*iso*thioureas (XIV) should therefore be particularly suitable starting materials in biguanide synthesis. This route, described in the patent literature (*154, 515, 516*), has been re-examined in detail by *Curd, Rose* et al. (*58, 133*). No desulphurising agent is required in this modification (*133*).

Aromatic members of this series (XIV; R = Aryl, R' = H or Aryl) reacted with alkylamines in methanol under a variety of conditions to afford variable yields (20—90%) of biguanides (*58, 377*). In many examples, however, cyanoguanidine was again the main product, its formation being apparently favoured by the presence of water (*58*).

Aliphatic S-alkylamidino*iso*thioureas (XIV; R = R' = Alk) and arylamines react readily (*143*) under conditions favouring the condensation of alkylcyanoguanidines with arylamines. This provides further evidence, though not proof, that the reaction involves intermediate alkylcyanoguanidines (X).

In the preparation of 1,5-disubstituted biguanides by this method, alcoholic ammonia at 100° in a closed vessel is used (*133*). Thiourea is an effective desulphurising agent (*133*).

The reaction between the pyrazole derivative (XV) and amines has given results of interest (*580*): when (XV) is refluxed in cyclohexylamine or morpholine ($Y = C_6H_{11}$ or N<⌒>O) (i.e. at ca. 130°) both the pyrazole- and the methylthio-groups are replaced, with formation of 1,4-disubstituted 5-phenylbiguanide (XVI; $Y = C_6H_{11}$ or N<⌒>O). At the lower boiling point (78°) of n-butylamine, the pyrazole-group is still replaced, but the methylthio-group remains unattacked, 1-(n-butylamidino)-2-S-methyl-3-phenylisothiourea (XVII) being obtained (*580*).

```
YNH·C·NH·C·NHPh  <--YNH2--  [pyrazol-1-yl]  SMe        --nBuNH2-->  nBuNH·C·NH·C=NPh
    ‖    ‖                        |         |                             ‖     |
    NH   NY                  HN=C·NH·C=NPh                               NH    SMe

   (XVI)                          (XV)                                  (XVII)
```

3. Mechanism

During their study of the synthesis of Paludrine-analogues, *Curd, Rose* et al. (*58*) observed that ammonolysis in ethanol of the S-methylisothiourea (XVIII; $R = p\text{-}ClC_6H_4$; Alk = isoPr) gave, instead of the expected Paludrine (XIX), the O-ethylisourea (XXI). Their interpretation of this result (postulating the intermediate formation of a carbodiimide (XX), and subsequent addition of ethanol) appears open to question, however, because carbodiimides are known to react with alcohols only with difficulty except in the presence of catalysts (*161*, *373*). In the absence of ethanol, aminolysis of XVIII by isopropylamine yields 1-p-chlorophenyl-4,5-diisopropylbiguanide as expected (*58*).

```
RNH·C·NH·C=NAlk  ---->  [RNH·C·N=C=NAlk]  --EtOH-->  RNH·C·NH·C=NAlk
    ‖    |                    ‖                          ‖     |
    NH   SMe                  NH                         NH    OEt
    ╳ (XVIII)                (XX)                        (XXI)
    ↓

RNH·C·NH·C·NHAlk  <--HgO--  RNH·C·NH2 + NH2CSNHAlk
    ‖    ‖                      ‖
    NH   NH                     NH
(XIX)
```

## D. Condensation of Guanidines and Thioureas

Under the influence of desulphurising agents, guanidines and thioureas combine to form biguanides (XIX), albeit in low yields (*122*, *515*). In their re-examination of this reaction, *Curd, Rose* et al. (*24*, *57*, *154*) synthesised 1,5-disubstituted biguanides; however, the yields did not exceeded 5%, irrespective of the desulphurising agent used. According to later claims (*633, 634*), improved yields are achieved in high-boiling non-polar solvents (e.g. toluene, xylene, anisole, di-n-butyl ether) at temperatures above 110°. 1,3-Disubstituted thioureas afforded 1,2,5-trisubstituted biguanides, again in low yields, by the mercuric oxide-ethanol procedure (*24*).

In the conventional extension (*24*) of this synthesis to S-alkylisothioureas, low yields of the expected biguanides were obtained. *Curd, Rose* et al. (*128, 131*) have improved this synthesis by performing the condensation under conditions which first convert the S-alkylisothiourea into the corresponding cyanamide (i.e. in non-hydroxylic organic media boiling above 100°, and in the presence of an alkali metal alkoxide).

## E. Ammonolysis and Aminolysis of Dithiobiurets

1-Aryldithiobiurets (*371*), on reaction with ammonia and a desulphurising agent, are reported to yield cyanoguanidines (*681*).

S-Alkyl-derivatives of dithiobiuret undergo aminolysis, as expected (*141, 144*). In the examples studied a desulphurising agent was needed to prevent the aminolysis stopping at the guanylthiourea stage. Details of the behaviour (including some anomalous results) of p-chlorophenyldithiobiuret (*208*), and its 4-S-methyl- (*208*) and 4-S-ethylderivative (*295*) in this group of reactions are on record.

That the reaction may proceed via the corresponding carbodiimide [cf. guanylthioureas (see Section III, C3)], is supported by the observation that interaction of 1-p-chlorophenyl-5-isopropyl-4-thiobiuret (XXII) and mercuric oxide in methanol gave 1-p-chlorophenyl-4-methyl-5-isopropyl-4-isobiuret (XXIII) (*141*):

$$\underset{\text{(XXII)}}{p\text{-}ClC_6H_4\cdot NH\cdot \underset{\underset{O}{\|}}{C}\cdot NH\cdot \underset{\underset{S}{\|}}{C}\cdot NHPr^i} \xrightarrow[\text{MeOH}]{\text{HgO}} \underset{\text{(XXIII)}}{p\text{-}ClC_6H_4\cdot NH\cdot \underset{\underset{O}{\|}}{C}\cdot NH\cdot \underset{\underset{OMe}{|}}{C}{=}NPr^i}$$

Further, isodithiobiurets of the type $ArNH\cdot CS\cdot N:C(SAlk)NAlk_2$ (lacking the necessary hydrogen atoms for carbodiimide formation) are not convertible into biguanides (*141*).

The sulphur atoms in dithiobiuret can be successively alkylated. It is therefore possible to synthesise 1,5-disubstituted biguanides carrying *unlike* substituents by the stepwise alkylation and replacement of the alkylthio-groups of dithiobiuret, as has been described (*295*) for Paludrine (XXVIII, $R = p\text{-}ClC_6H_4$).

$$\underset{\text{(XXIV)}}{\underset{\text{S}\quad\ \text{S}}{\text{RNH}\cdot\text{C}\cdot\text{NH}\cdot\text{CNH}_2}} \xrightarrow{\text{EtI}} \underset{\text{(XXV)}}{\underset{\text{SEt}\quad\ \text{S}}{\text{RN}=\text{C}\cdot\text{NH}\cdot\text{C}\cdot\text{NH}_2}} \xrightarrow{\text{NH}_3} \underset{\text{(XXVI)}}{\underset{\text{NH}\quad\ \text{S}}{\text{RNH}\cdot\text{C}\cdot\text{NH}\cdot\text{C}\cdot\text{NH}_2}}$$

$$\xrightarrow{\text{EtI}} \underset{\text{(XXVII)}}{\underset{\text{NH}\quad\ \text{SEt}}{\text{RNH}\cdot\text{C}\cdot\text{N}=\text{C}\cdot\text{NH}_2}} \xrightarrow{\text{isoPrNH}_2} \underset{\text{(XXVIII)}}{\underset{\text{NH}\quad\ \text{NH}}{\text{RNH}\cdot\text{C}\cdot\text{NH}\cdot\text{C}\cdot\text{NHisoPr}}}$$

This attractive synthesis has the serious disadvantage that with the exception of the first stage, the yields (of XXVI to XXVIII) are low throughout (15%, 26%, and 10% respectively). According to the authors (*295*), the final compound (XXVIII) is in fact an isomeric form of Paludrine. This is discussed in Section III N.

## F. Ammonolysis of O-Alkylamidinoisoureas

Yet another similar route to biguanides utilises the ammonolysis and aminolysis of O-alkylamidino*iso*ureas. The use of this group of reactions, first briefly referred to in the patent literature (*516*), was developed by *Curd* and his co-workers who first published (*138, 139*) adequate experimental details; it was later patented in Japan (*353, 354*).

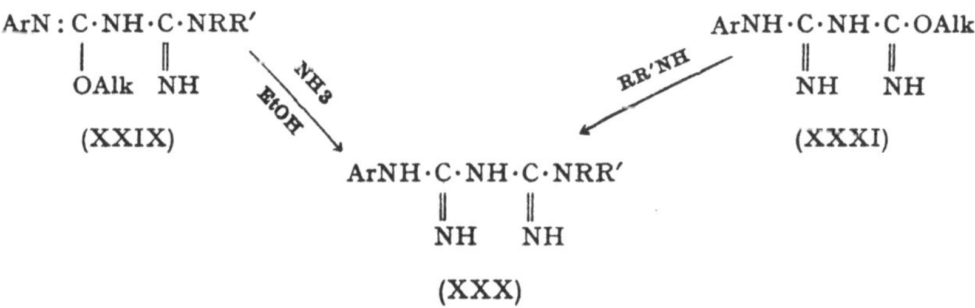


## G. Condensation of Guanidines and O-Alkylisoureas

The condensation of O-alkylisoureas and guanidines (R = R' = Alk) was also patented (*129, 130*) by *Curd* and his co-workers.

$$\underset{\text{NH}}{\text{ArNH}\cdot\text{C—OAlk}} + \underset{\text{NH}}{\text{N}_2\text{N}\cdot\text{C}\cdot\text{NR R}'} \rightarrow \underset{\text{NH}\quad\ \text{NH}}{\text{ArNH}\cdot\text{C}\cdot\text{NH}\cdot\text{C}\cdot\text{NRR}'} + \text{EtOH}$$

The interaction (*578*) of O-methylisourea and N-methylglycine occurs normally at low temperatures to yield the expected creatine (XXXII) (21%). At 40—50° the production of 1-methyl-1-biguanide acetic acid (XXXIII) in low yield has been reported, but the mechanism of its formation ist not clear.

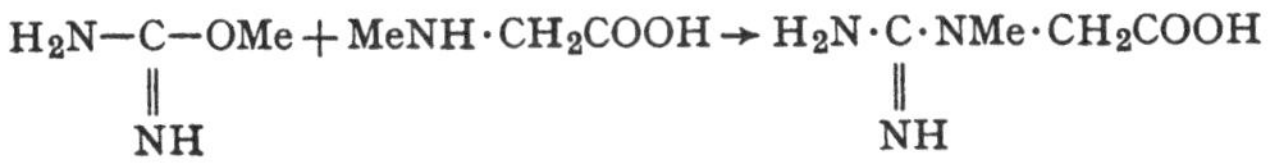

$$H_2N{-}\underset{\underset{NH}{\|}}{C}{-}OMe + MeNH{\cdot}CH_2COOH \rightarrow H_2N{\cdot}\underset{\underset{NH}{\|}}{C}{\cdot}NMe{\cdot}CH_2COOH$$

(XXXII)

$$H_2N{\cdot}\underset{\underset{NH}{\|}}{C}{\cdot}NH{\cdot}\underset{\underset{NH}{\|}}{C}{\cdot}NMe{\cdot}CH_2COOH$$

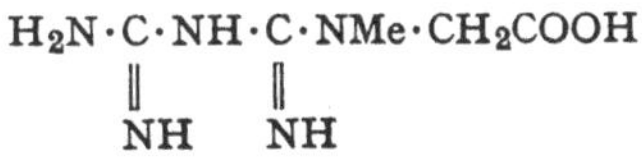

(XXXIII)

## H. Addition of Cyanamides to Guanidines

An inspection of the structure of biguanides (XXXIV)

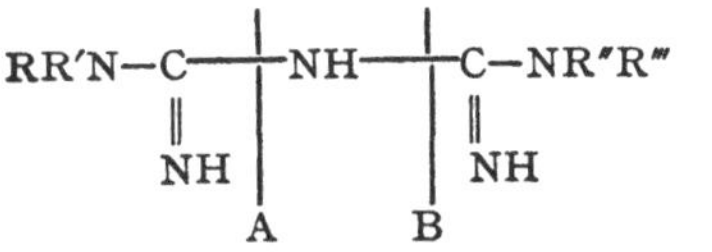

$$RR'N{-}\underset{\underset{NH}{\|}}{C}\;\big|_A\;{-}NH{-}\;\big|_B\;\underset{\underset{NH}{\|}}{C}{-}NR''R'''$$

(XXXIV)

shows that the molecule may be built up theoretically from the fragments indicated by the lines (A) or (B). The second alternative, i.e. the synthesis from amines and cyanoguanidines, is the subject of Section III B. The former approach, i.e. the condensation of guanidines and cyanamides, was described (*524*) as early as 1879, but the yields reported were minute. Later examples, without experimental details, were recorded in the patent literature (*515, 516*) and a single additional example was given in an Indian publication (*36*). The reaction has been re-investigated in great detail by *Curd*, *Rose* and their group (*5, 6, 7, 153, 154*). Whilst their work was awaiting publication (*5*), a synthesis of Paludrine by fusion of the arylguanidine and alkylcyanamide was described by another group of investigators (*302*).

The suitability of dialkylcyanamides was first examined (*5, 6*). In boiling butanol, pentanol or toluene, biguanides were formed, the best yields being obtainable in the last solvent (*192, 632*). The synthesis was also effected by fusion of the reactants at 130°. The rate of formation of the biguanide was followed by means of the sparingly soluble biguanide-copper complex: maximum biguanide formation occurs after 15—30 minutes from mono-, and after 2—3 hours from di-alkylcyanamides. The reaction, though successful with arylcyanamides (*5, 6, 7, 153, 154*), is apparently not generally applicable to disubstituted cyanamides (*5*) and is not generally promoted by using part of the guanidine in the form of its hydrochloride (*5*).

Although monoalkylcyanamides polymerise readily at high temperatures (*412*), they afford biguanides in yields not inferior to those from the more stable dialkylcyanamides. This favourable result may be due in part to the greater reaction rates of the monoalkylcyanamides.

The preparation and subsequent pyrolysis of a monoalkylcyanamide salt of a guanidine has been suggested (*50*) as a route to biguanides.

With N,N'-disubstituted guanidines the problem arises whether 1,2,5- (XXXV) or 1,3,5-trisubstituted biguanides (XXXVI) are produced, or a mixture of both. Relevant experiments (*5*) show that the former (XXXV) are in fact obtained to the exclusion of the latter.

$$\underset{\text{(XXXVI)}}{RNH\cdot\underset{\|}{\overset{}{C}}(=NH)\cdot NR'\cdot C(=NH)\cdot NHR''} \leftarrow RNH\cdot C(=NH)\cdot NHR' + R''NHCN \rightarrow \underset{\text{(XXXV)}}{RNH\cdot C(=NR')\cdot NH\cdot C(=NH)\cdot NHR''}$$

The interaction of p-methoxybenzylideneaminoguanidine (XXXVII; R = p-MeO·$C_6H_4$·) and dimethylcyanamide (*580*) gave 17% yields of alleged 1-dimethyl-3-(p-methoxybenzylideneamino)biguanide (XXXVIII). The possible condensation at the free amidino-grouping (of XXXVII) was not considered by the author.

$$\underset{\text{(XXXVII)}}{RCH{=}N\cdot NH\cdot C(=NH)\cdot NH_2} + Me_2N\cdot CN \rightarrow \underset{\text{(XXXVIII)}}{Me_2N\cdot C(=NH)\cdot N(N{=}CHR)\cdot C(=NH)\cdot NH_2}$$

## I. Interaction of Cyanamides and Amines

### 1. By Initial Cyanamide Dimerisation

An ingenious use of monoalkyl- or alkoxyalkylcyanamides in the synthesis of biguanides utilises (*201*) their rapid dimerisation at elevated temperatures to cyanoguanidines (XXXIX), which are subsequently treated with hydroxyalkylamines: any unchanged cyanamide would yield a guanidine;

the fact that none is found indicates that dimerisation is complete. This reaction provides an exceptionally easy route to 1,3,5-trisubstituted biguanides (XL).

$$\mathrm{R{\cdot}NHCN} \xrightarrow{130^\circ} \underset{\text{(XXXIX)}}{\mathrm{RNH{\cdot}\underset{\underset{NH}{\|}}{C}{\cdot}NR{\cdot}CN}} \xrightarrow{\mathrm{R'R''NH}} \underset{\text{(XL)}}{\mathrm{RNH{\cdot}\underset{\underset{NH}{\|}}{C}{\cdot}NR{\cdot}\underset{\underset{NH}{\|}}{C}{\cdot}NR'R''}}$$

### 2. From Aniline and Cyanamide

The reaction between three moles of aniline hydrochloride and one of cyanamide is stated (*409*) to furnish triphenylbiguanide, but no structure for the product has been given.

### 3. As Secondary Products

The formation of a biguanide as a by-product in a condensation of a cyanamide and amine (*225*) (which would normally be expected to yield a guanidine) and as main product in the reaction of glucosamine and cyanamide (*437*) has been noted. In the former example, the small amount of biguanide probably arose by further reaction of the primary guanidine (which formed the main product) with cyanamide. In the second example the biguanide may be the result of the interaction of the amine and intermediate cyanoguanidine.

## J. From Dicyanimide and Amines

Another variation of the synthesis of biguanides utilising the condensation of cyano-compounds and amines is the reaction of dicyanimide and two moles of amine (*457, 465, 581, 582*). In practice (*457*) a dicyanimide metal salt is fused, or heated in water or in an organic solvent with an amine salt, preferably between 90 and 130°. Condensation may also be effected using free dicyanimide in aqueous acid media (*581*).

$$2\,\mathrm{RR'NH} + \mathrm{NC{\cdot}NH{\cdot}CN} \rightarrow \mathrm{RR'N{\cdot}\underset{\underset{NH}{\|}}{C}{\cdot}NH{\cdot}\underset{\underset{NH}{\|}}{C}{\cdot}NRR'}$$

The method is normally limited to the production of symmetrically substituted biguanides; by the stepwise addition of unlike amines, biguanides bearing different 1- and 5-substituents (e.g. iso$C_6H_{13}$NH·

·C( : NH)·NH·C( : NH)·NEtPh) (*534*) are obtainable. It will be noted that this variation of the synthesis approaches the general route dealt with in Section III B.

## K. Addition of Carbodiimides to Guanidines

1,2-Disubstituted biguanides and more highly substituted analogues may be synthesised by the addition of carbodiimides to guanidines (*122*).

Reaction is terminated at the desired monoaddition-stage by using free guanidine (base), and acetone as solvent (*376, 377*). Guanidine salts and carbodiimides in dimethylformamide, in spite of the presence of excess of the former, yield 1,2,6-trisubstituted isomelamines (XLIII), presumably by loss of arylamine from the intermediate labile triguanide (XLII) (*375*).

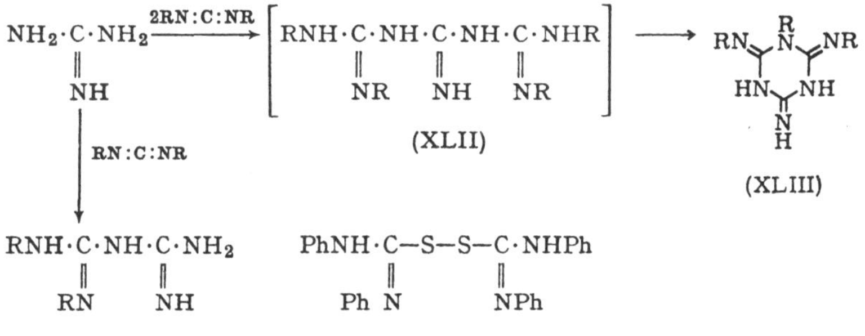

$$NH_2{\cdot}C({:}NH){\cdot}NH_2 \xrightarrow{2RN:C:NR} \left[RNH{\cdot}C({:}NR){\cdot}NH{\cdot}C({:}NH){\cdot}NH{\cdot}C({:}NR){\cdot}NHR\right] \text{(XLII)} \longrightarrow \text{(XLIII)}$$

$$NH_2{\cdot}C({:}NH){\cdot}NH_2 \xrightarrow{RN:C:NR} RNH{\cdot}C({:}NR){\cdot}NH{\cdot}C({:}NH){\cdot}NH_2 \text{ (XLI)}$$

(XLIII): six-membered ring RN=C–N(R)–C(=NR)–NH–C(=NH)–NH–

$$PhNH{\cdot}C({:}NPh){-}S{-}S{-}C({:}NPh){\cdot}NHPh \text{ (XLIV)}$$

sym-Triphenylguanidine and diphenylcarbodiimide similarly yield sym-pentaphenylbiguanide (*420*). The thermal decomposition of tetraphenyldithioformamidine (XLIV), giving sym-pentaphenylbiguanide as one of the products, has been interpreted in terms of the above reaction. The primary decomposition products are, inter alia, diphenylcarbodiimide and sym-triphenylguanidine, which then react as above (*323*).

The addition of carbodiimides to N,N'-diaminoguanidine is yet another extension of this general synthesis. Although it is the hydrazino-group of diaminoguanidine that participates more rapidly in this addition-process, the central imino-group becomes the reactive centre when the hydrazino-functions are suitably blocked, e.g. by hydrazone-formation. Thus, the acetone dihydrazone of N,N'diaminoguanidine reacts with diphenylcarbodiimide yielding 3-(N,N'-diphenylamidino)-1,2-di(isopropylideneamino)guanidine [i. e. PhNH·C( : NPh)·N : C(NHN : $CMe_2)_2$] which is in fact a substituted 1,2-diaminobiguanide (*374*). The compound readily undergoes cyclisation to 1,2,4-triazoles (*374*).

## L. Synthesis from Aminomagnesium Halides

Biguanides are accessible by a Grignard reaction of guanidinomagnesium halide with a substituted cyanamide, and hydrolysis of the resulting complex (*59–61*). The method is of considerable theoretical interest but so far only of limited practical importance.

The required Grignard reagents are formed from the substituted guanidines and ethyl magnesium halide in ether (*59*), with simultaneous evolution of ethane. Their assigned structure (XLV) accounts satisfactorily for their reactions, although their true structure will obviously be essentially ionic and exhibit resonance effects.

The interaction of guanidinomagnesium halides (XLV) with disubstituted cyanamides (XLVI), for example, affords substituted biguanides (XLVIII) (*59*). The variation in yields (7 to 48%) is attributed (*59*) to differences in solubility in ether of the guanidinomagnesium halides and the resultant complexes, and to the electronic effects of the substituents.

Monosubstituted cyanamides require two equivalents of guanidinomagnesium halide to allow for replacement of the acidic cyanamide hydrogen by the MgI moeity (*59*). The reaction proceeds in low yield and fails with monoarylcyanamides.

Aminomagnesium halides (XLIX) may also serve as a source of the same substituted biguanides (XLVIII) by reaction with substituted cyanoguanidines (L), but yields are again low and variable. The possible mechanism of this group of reactions has been discussed (*59*), but the studies are as yet incomplete.

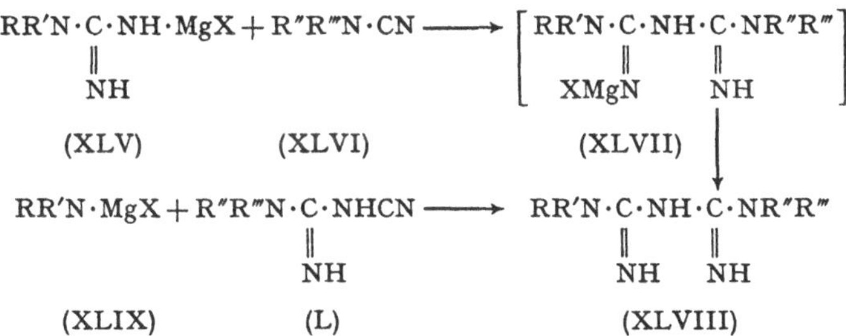


## M. Miscellaneous Syntheses

### 1. From Biguanide and Aromatic Amines

A patented (*113*) preparation of monoarylbiguanides consists of prolonged refluxing of an aromatic amine with biguanide in 10–15% hydrochloric acid. Although biguanides are stable in boiling hydrochloric acid of this strength for short periods (see Section VII C), the claim must be regarded with reserve in view of the different reaction of substituted biguanides and amines which has been reported (*166*) (cf. Section VII, C).

2. From Diphenylcarbodiimide

sym-Pentaphenylbiguanide [PhNH·C(:NPh)NPh·C(:NPh)NHPh] arises in the interaction of equimolar amounts of diphenylcarbodiimide and triethylphosphine (*327*).

3. From Suitable Chlorine-Containing Compounds

A single report (*315*) claims a synthesis of biguanides which utilises the replacement of a chlorine atom by an amino-group in suitable guanylchloroformamidines [e.g. $R_2N\cdot C(:NH)\cdot N{=}CCl\cdot NR_2$] (*314*). The difficulty of preparing the starting materials (*314*) would appear to exclude the widespread use of this synthesis.

4. From trans-1-Amino-2-hydroxycyclohexane

An unexpected formation of a biguanide occurred (*754*) in the reaction of trans-1-amino-2-hydroxycyclohexane (LI) with cyanogen bromide in the presence of sodium acetate. In addition to the expected trans-2-amino-3a,4,5,6,7,7a-hexahydro-benzoxazole (LII), traces of its trimer having the biguanide structure (LIII) were obtained.

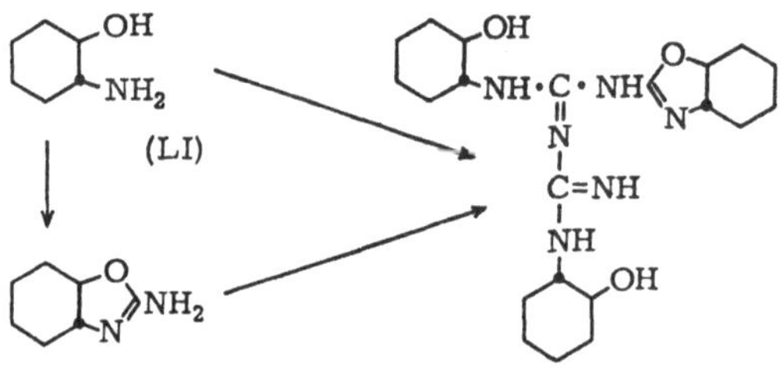

(LI)

(LII)

(LIII)

This trimer was then synthesised by refluxing the benzoxazole (LII) with toluene-p-sulphonic acid in benzene. In contrast, the cis-isomer of (LII) is stable under these conditions. The mechanism of formation of the trimer (LIII) was discussed (*754*).

5. From Selected Triazines

Biguanides are also formed by the breakdown of certain triazines. Thus alcoholic hydrochloric acid converts 2-dichloromethyl-6-phenylguanamine (LIV) into phenylbiguanide (LVI, R=Ph) (*443*), and aqueous ethanol cleaves 1-p-nitrophenyl-1,2-dihydro-4,6-diamino-2,2-dimethyl-s-triazine (LV) to p-nitrophenylbiguanide (LVI, R = p-$NO_2C_6H_4$.) under surprisingly gentle conditions (*441*). The readiness of the latter reaction has been ascribed (*441*) to the effect of the electron-attracting nitro-group which reduces the basicity of the nitrogen atom at $N^1$. As, however, in both cases a biguanide is used to synthesise the triazines concerned, the practical value of these reactions is limited.

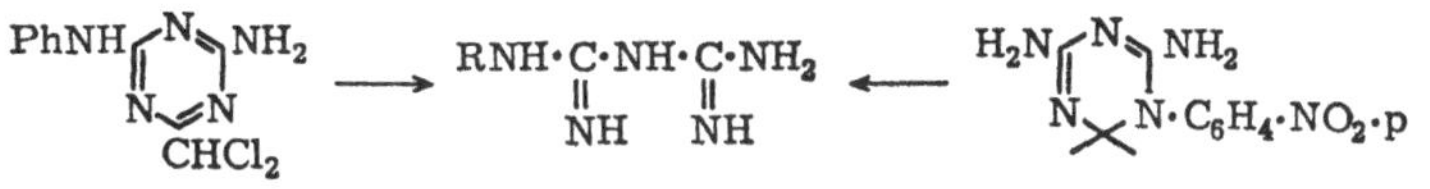

PhNH-triazine(NH$_2$)(CHCl$_2$) ⟶ RNH·C(=NH)·NH·C(=NH)·NH$_2$ ⟵ H$_2$N-triazine(NH$_2$)(N·$C_6H_4$·$NO_2$·p)(Me$_2$)

(LIV) (LVI) (LV)

It might be noted here that p-nitrophenylbiguanide is remarkable in that it can be synthesised (*441*) from p-nitroaniline and cyanoguanidine at the very low temperature of boiling acetone in excellent yield and purity.

## N. Isomeric Forms of Biguanides

Two reports in the literature have dealt with alleged isomers of Paludrine. A sample of this compound, obtained in 1947 by Russian workers (*45*) by the standard synthesis had m.p. 129° (from benzene) and thus appeared to be identical with the material of *Curd, Rose* et al. (*143*). Prolonged drying and seeding allegedly converted this "labile form" (m.p. 129°) into a "stable form", m.p. 144° (*45*).

A second report of this alleged phenomenon appeared in 1952 (*295*). Specimens of Paludrine synthesised from 1-p-chlorophenyldithiobiuret (see Section III E) had m.p. 201—202° (instead of 129°) (*143*), and were regarded as an isomer of Paludrine. It is interesting in this context that successive replacement of sulphur atoms in the opposite order (i.e. 4 and 2) by isopropylamine and ammonia in 1-p-chlorophenyldithiobiuret yields (*296*) authentic Paludrine, m.p. 129°. There is no doubt that structural assignments to these alleged isomers remain to be confirmed.

# IV. Formation of Triguanides from Biguanides

Following the discovery of the antimalarial properties of Paludrine (*148*), an obvious step in the search for active compounds was the synthesis of the comparable triguanides (LVII). Several attempts by Indian investigators with this object in view appear to have been unsuccessful (*304*).

Thus, the interaction of various monosubstituted biguanides and cyanamides under a variety of conditions gave, in some cases, 2,4-disubstituted melamines (LVIII) (and sometimes melamine itself), that may have arisen from primary intermediate triguanides (LVII). The attempted condensations of two moles of a cyanamide with one of guanidine, or of equimolar quantities of a cyanoguanidine and guanidine also failed, giving in each case unidentified products that were not the desired triguanide.

$$RNH{\cdot}C({=}NH){\cdot}NH{\cdot}C({=}NH){\cdot}NH_2 + R'NCS \longrightarrow [RNH{\cdot}C({=}NH){\cdot}NH{\cdot}C({=}NH){\cdot}NH{\cdot}C({=}S){\cdot}NHR'] \xrightarrow{-H_2S} \text{(LX)}$$

(LIX)

$$RNH{\cdot}C({=}NH){\cdot}NH{\cdot}C({=}NH){\cdot}NH_2 + R'NHCN \longrightarrow [RNH{\cdot}C({=}NH){\cdot}NH{\cdot}C({=}NH){\cdot}NH{\cdot}C({=}NH){\cdot}NHR'] \longrightarrow \text{(LVIII)}$$

(LVII)

$$RNHCN + NH_2{\cdot}C({=}NH){\cdot}NH_2 + RNHCN \not\longrightarrow \text{(LVII)}$$

$$RNH{\cdot}C({=}NH){\cdot}NHCN + NH_2{\cdot}C({=}NH){\cdot}NHR' \not\longrightarrow \text{(LVII)}$$

(LX): HN, R–N, NR', HN, NH, NH (ring)

(LVIII): RNH, N, NHR', N, N, $NH_2$ (ring)

A synthesis of triguanides, claimed by workers (*297*) of the same laboratory, consists of the condensation of substituted biguanides with isothiocyanates to the thioureido-biguanides (LIX), and their subsequent ammonolysis, in presence of mercuric oxide, to (LVII). However, it has since been conclusively shown (*377*) that under the specified conditions the products of this reaction are in fact 1,2-disubstituted isomelamines (LX); they may arise from the primary open-chain adducts (LIX) by immediate cyclisation with loss of hydrogen sulphide. The original claims (*297*) are therefore regarded as unsubstantiated, as are the results of a later paper (*557*) employing the alleged synthesis (see also Section III B).

A preliminary report (*461*) has claimed a synthesis of 1,1-dialkyltriguanides by fusion at 180° of biguanide salts with dialkylcyanamides. In the absence of fuller details, this report is regarded with reserve, in view of the established instability (*377*) of triguanides (*304*) under the conditions specified for this synthesis.

## V. Formation of Tetraguanides

Speculations on the existence of tetraguanides have appeared (*478*) in the patent literature. The production of biguanides by fusion of cyanoguanidine with amine hydrochlorides is often attended by evolution of ammonia. This observation prompted the statement "that the reaction often proceeds . . . beyond the formation of the simple biguanide radical . . . Possibly two biguanide radicals lose ammonia with formation of a polyguanide radical such as . . . a tetraguanide." In view of the uncertainty concerning the existence of triguanides and the comparable tetruret (*371*) the production of tetraguanides at these elevated temperatures must be regarded as very questionable.

## VI. Physical Properties

### A. General

Biguanides, no matter how highly substituted (*57*), are strong bases that form stable salts with inorganic and organic acids (e.g. picric (*332*), picrolonic (*759*), tannic (*332, 712*), mesotartaric, mesoxalic, tartronic (*723*)). Being diacid bases, they normally yield dipicrates (*604*) but monopicrates are also formed, particularly with increasing bulk of their $N^1$-substituent.

In the dihydrochloride (*602*), the second molecule of acid is not firmly bound because of protonation of the second basic site by internal hydrogen bond formation. This would suggest that in the dipicrate, one mole of picric acid participates in salt formation, and that the second mole of picric acid is added to the hydrogen-bonded cation (LXI) forming a molecular complex. The existence of a monopicrate might thus serve to point out structures with steric inhibition of hydrogen bonding.

RR'N–C(=NH)–N=C(–NH)–$NH_2$, with HN···H–NH hydrogen bond (LXI)

Heptamethylbiguanide catalyses the addition of phenyl isocyanate to alcohols more powerfully than do other amines, a property that is ascribed to its high basic strength (*211*).

Salts of substituted biguanides with N-acylsulphonamides, of general structure $[RSO_2N \cdot COR']^{\ominus}[R''NH_2C(:NH)NH \cdot C(:NH)NH_2]^{\oplus}$ have been prepared (*380*), as have addition complexes with 2-hydroxy-4,6-dimethylpyrimidine (*411*).

The 1,5-disubstituted biguanides have been described in detail (*148*). They are generally low-melting solids which are sparingly soluble in cold water. The lower members are sufficiently soluble to give strongly alkaline solutions. They rapidly absorb atmospheric carbon dioxide and are best characterised as their stable and crystalline acid salts (*148, 149, 152, 153, 232*).

The crystal structure of 1-p-chlorophenyl-5-isopropylbiguanide hydrochloride ("Paludrine") and some analogues has been examined by X-ray diffraction methods (*86*) (see page 380). The organic cation is roughly U-shaped, and the chloride ion is coordinated to six nitrogen atoms. The biguanide part of the cation consists of two planes, intersecting at 58.9°, each containing a $CN_3$-group, with a mean C—N length of 1.328 Å.

The unit cell dimensions and the space-group of biguanide acid sulphate are (*440*) a 11.507; b 20.446; c 7.203 Å, and Pbca. Data are also available for ruthenium biguanide sulphate (*287*) and copper biguanide chloride (*288*).

## B. Spectra and Dissociation Constants

### 1. Introduction

A knowledge of the acid strengths and absorption properties of biguanides has provided much of the basis for discussing the more detailed structure of the biguanide molecule. One of the earlier comprehensive studies (*227*) of the ultraviolet absorption properties of biguanides was aimed at correlating their structure and antimalarial activity. Although this objective was not attained, the measurements provided much useful information, including, for example, data for calculating dissociation constants of biguanide and its derivatives (see Table 1).

### 2. Acid Dissociation Constants

Biguanides usually behave as mono- and di-acid bases, combining with $H^+$ ions to form the conjugate acid, but may also act as acids. The acid dissociation constants of many biguanides have been measured, and are collected in Tables 1 and 2.

$$H_2N\cdot\underset{\|}{\overset{}{C}}(=NH)\cdot NH\cdot C(=NH)\cdot\overset{+}{N}H_3 \xleftarrow{H^+} H_2N\cdot C(=NH)\cdot NH\cdot C(=NH)\cdot NH_2 \rightarrow H_2N\cdot C(=NH)\cdot NH\cdot C(=N^-)\cdot NH_2 + H^+$$

Table 1. *Ultraviolet spectra and dissociation constants* $RR_1N \cdot C(=NR_2) \cdot NH \cdot C(=NR_3) \cdot NHR_4$

| R | $R_1$ | $R_2$ | $R_3$ | $R_4$ | $pK_1$ | $pK_2$ | $\lambda_{max}B$ mμ | $\varepsilon_{max}B$ | $\lambda_{max}BH^+$ mμ | $\varepsilon_{max}BH^+$ | $\lambda_{max}BH_2^{++}$ mμ | $\varepsilon_{max}BH_2^{++}$ | Ref. |
|---|---|---|---|---|---|---|---|---|---|---|---|---|---|
| | | | | | | | | | [235] | [13,500] | <225 | — | |
| $C_6H_4Cl$ | H | H | H | H | 10.4 | 2.2 | 239 | 13,080 | 253 | 14,800 | 245 | 8,600 | (*227*) |
| | | | | | | | | | | | <220 | — | |
| $C_6H_5$ | H | H | H | $Pr^i$ | 11.5 | 1.9 | 236 | 12,700 | 245 | 13,300 | [245] | [5,500] | (*227*) |
| | | | | | | | | | 233 | 14,000 | <225 | — | |
| $C_6H_4Cl$ | H | H | H | $Pr^i$ | 10.4 | 2.3 | 240 | 14,470 | 253 | 14,750 | [244] | [8,850] | (*227*) |
| | | | | | | | | | <230 | — | 230 | — | (*227*) |
| $C_6H_4Br$ | H | H | H | $Pr^i$ | 11.4 | 2.2 | 246 | 14,400 | 256 | 15,100 | 247 | 10,000 | (*227*) |
| | | | | | | | | | ~237 | 14,800 | <230 | — | |
| $C_6H_4I$ | H | H | H | $Pr^i$ | 11.0 | 2.2 | 248 | 15,800 | 260 | 16,200 | ~250 | 10,150 | (*227*) |
| $C_6H_4Cl$ | Me | H | H | $Pr^i$ | 12.8 | 2.4 | 243 | 13,500 | 243 | 15,600 | <225 | — | (*227*) |
| | | | | | | | 237 | 13,000 | 237 | | <225 | — | |
| $C_6H_4Cl$ | H | Me | H | $Pr^i$ | 11.8 | 2.3 | [260] | [10,650] | 236 | 18,700 | 252 | 9,900 | (*227*) |
| | | | | | | | | | <220 | | | | |
| $C_6H_4Cl$ | H | H | Me | $Pr^i$ | 12.2 | 2.0 | 240 | 13,600 | 250 | 13,800 | <220 | — | (*227*) |
| H | H | H | H | H | 12.8 | 3.1 | 231 | 9,500 | 232 | 12,500 | <220 | | (*227*) |
| H | H | H | H | H | 13.3 | 3.2 | 230 | 8,900 | 230 | 12,360 | <220 | | (*322*) |

Table 2. *Dissociation constants of biguanides*

| Compound | $pK_1$ | $pK_2$ | Ref. |
|---|---|---|---|
| Biguanide | 11.51 | 2.94 | *(177)* |
| | 11.52 | 2.93 | *(569)* |
| | 11.50 | 2.93 | *(569)* |
| | 11.49 | 2.95 | *(37)* |
| | 11.52 | 2.94 | *(37)* |
| | > 4.00 | 2.86 | *(544)* |
| Methylbiguanide | 11.44 | 3.00 | *(569)* |
| | 11.41 | 2.99 | *(569)* |
| Ethylbiguanide | 11.47 | 3.08 | *(569)* |
| Dimethylbiguanide | 11.52 | 2.77 | *(569)* |
| | 11.53 | 2.73 | *(569)* |
| 1,2,5-Trimethylbiguanide | 17.2[a] | | *(666)* |
| 1,2,4,5-Tetramethylbiguanide | 17.2[a] | | *(666)* |
| Methoxyethylbiguanide | 11.47[a] | | *(301)* |
| Diethylbiguanide | 11.68 | 2.53 | *(569)* |
| | 11.64 | 2.56 | *(569)* |
| n-Propylbiguanide | 11.35 | 3.10 | *(194)* |
| Propylbiguanide | 11.35[a] | | *(301)* |
| Methoxypropylbiguanide | 11.40[a] | | *(301)* |
| iso-Propylbiguanide | 11.35 | 3.10 | *(194)* |
| n-Butylbiguanide | 11.28 | 2.92 | *(194)* |
| n-Hexylbiguanide | 11.44 | 3.30 | *(194)* |
| Cyclohexylbiguanide | 11.39 | 3.20 | *(194)* |
| Ethylhexylbiguanide | 11.40 | 3.10 | *(194)* |
| Propanolbiguanide | 11.50[a] | | *(301)* |
| Ethanolbiguanide | 11.53[a] | | *(301)* |
| Benzylbiguanide | 11.25 | 2.70 | *(194)* |
| p-Phenetylbiguanide | 10.82 | 2.70 | *(194)* |
| Phenylbiguanide | 10.72 | 2.14 | *(177)* |
| | 10.72 | 2.16 | *(569)* |
| | 10.76 | 2.13 | *(569)* |
| | 10.70 | 2.16 | *(37)* |
| | 10.72 | 2.16 | *(37)* |
| p-Tolylbiguanide | 10.84 | 2.60 | *(194)* |
| Naphthylbiguanide | 10.24 | 2.05 | *(569)* |
| 1-(2′-Anthryl)biguanide | 10.4 | | *(558)* |
| 1-(3′-Acridyl)biguanide | [b] | | *(558)* |
| 1-(6′-Quinolyl)biguanide | 10.7[a] | | *(558)* |
| 1-(8′-Hydroxyquinol-5′-yl)biguanide | 11.8[a] | | *(9)* |

[a] Although not specified in the papers, the values are listed here as first acid dissociation constants, in view of their numerical values.

[b] A pK value was not obtained for this compound because of its strong tendency to form micelles, even in dilute solutions.

Table 3. *Dissociation constants of dibiguanides*

| Compound | $pK_1$ | $pK_2$ | $pK_3$ | $pK_4$ | Ref. |
|---|---|---|---|---|---|
| Ethylenedibiguanide | 11.76 | 11.34 | 2.88 | 1.74 | (*569*) |
| | 11.83 | 11.28 | 2.90 | 1.76 | (*569*) |
| Phenylenedibiguanide | 11.39 | 9.87 | 2.15 | 1.41 | (*569*) |
| | 11.35 | 9.80 | 2.18 | 1.38 | (*569*) |
| Hexamethylenedibiguanide | 11.54 | 10.75 | 3.49 | 2.28 | (*194*) |
| | 11.52 | 10.76 | 3.51 | 2.36 | (*194*) |

The $pK_1$ values show clearly that biguanide, and its mono- and dialkyl-homologues are stronger bases than are arylbiguanides. Similar observations are made in the dibiguanide series (see Table 3).

The pK values computed from ultraviolet absorption measurements (*227*) confirm that biguanides are invariably di-acid bases. Since the parent compound, biguanide, is a strong base, a reported $pK_1$ value of 12.8 (*227*) must be regarded with some reserve. Its $pK_2$ (3.1) is lower than might be expected for the singly charged cation (LXII), the $pK_2$ of the structurally comparable malondiamidine being 9.0. The singly charged cations of both these compounds are possibly stabilised by hydrogen bond resonance (LXIII a,b: R=NH or $CH_2$) (see also ref. *206*). In biguanide, additional ionic species (LXIV a,b; LXV) are also possible. In (LXV) all the four terminal amino-groups are approximately equivalent; the cationic charge will be distributed among them and the ion will be stabilised by a high resonance energy. However, the structure of the singly charged cation can clearly not be assigned from a study of dissociation constants alone.

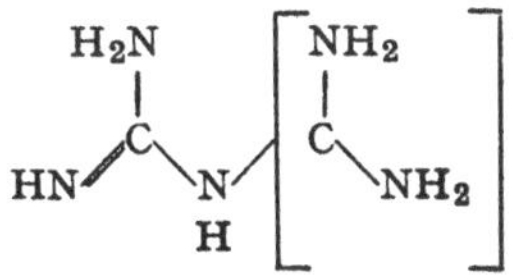

(LXII) 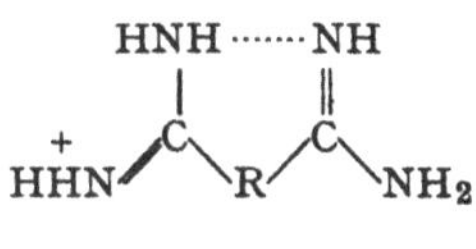 (LXIII a) 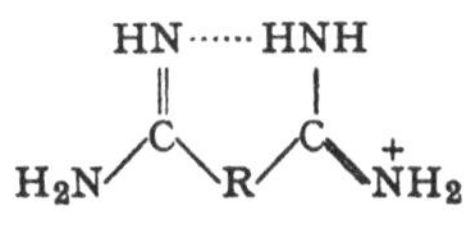 (LXIII b)

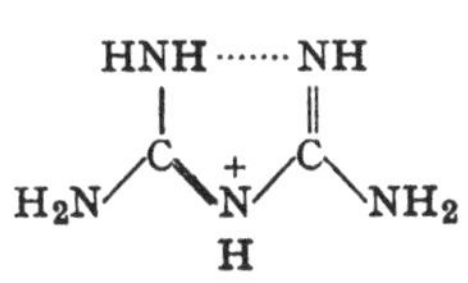

(LXIV a) 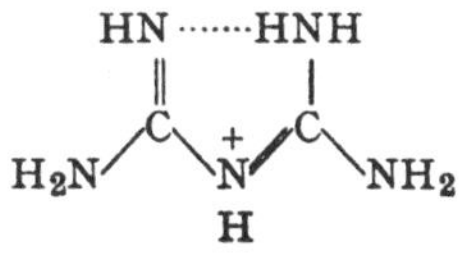 (LXIV b) 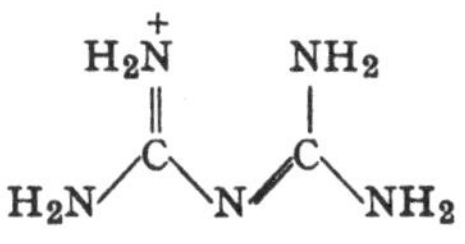 (LXV)

### 3. Ultraviolet Absorption Spectra

Attempts to gain further insight into the more detailed structure of biguanides in general, and of Paludrine in particular, were made by studies of their ultraviolet spectra, including those of simpler molecules, which may be regarded as the component parts of 1-aryl-5-alkylbiguanides (*227*), but the succes of this approach was at first limited (*227*).

The spectrum and dissociation constants of biguanide were later (*322*) redetermined and substantially confirmed, values being recorded for the neutral molecule (at pH 14.5 and above) and for the singly and doubly charged positive ions (at pH 5 to 10 and pH $<1.5$, respectively) (see Table 1). Single protonation of biguanide does not displace its absorption maximum but only decreases its intensity. Addition of a second proton excludes the existence of a conjugated resonating system, and displaces the absorption band into the vacuum ultraviolet region (see also ref. *687*). From these observations, biguanide and its ions were represented thus:

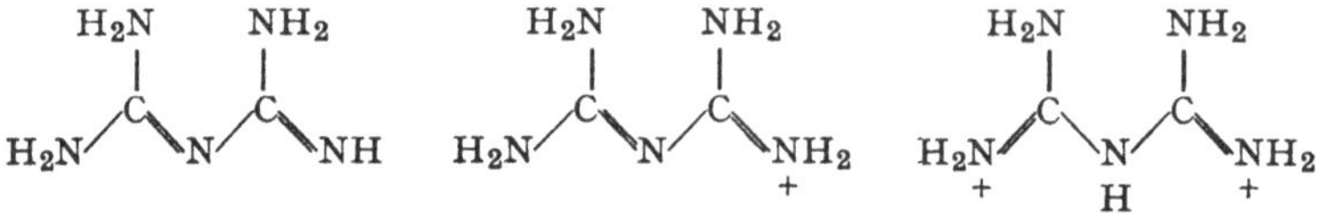

$H_2N{-}C({=}NH_2\ldots)$ structures:

$$H_2N{-}C(NH_2){=}N{-}C(NH_2){=}NH \qquad H_2N{-}C(NH_2){=}N{-}C(NH_2){=}\overset{+}{N}H_2 \qquad H_2\overset{+}{N}{=}C(NH_2){-}NH{-}C(NH_2){=}\overset{+}{N}H_2$$

A further comprehensive study, but limited to 1-(β-phenethyl)-biguanide has been undertaken by *Shapiro* et al. (*602*). On the basis of their experimental findings (Table 4) they summarised (*602*) the behaviour of this compound at various hydrogen ion concentrations by equations (i) – (vii) ($B = PhCH_2CH_2{\cdot}NH{\cdot}C(:NH)NH{\cdot}C(:NH)NH_2$).

$$BH^+Cl^-(H_2O) \longrightarrow BH^+ + Cl^- \qquad (i)$$

$$BH^+Cl^-(HCl) \longrightarrow BH_2^{++} + 2\,Cl^- \qquad (ii)$$

$$BH_2^{++}Cl_2^{--}(H_2O) \longrightarrow BH^+ + 2\,Cl^- + H^+ \qquad (iii)$$

$$B(H_2O) \longrightarrow BH^+ + OH^- \qquad (iv)$$

$$BH^+(NaOH) + OH^- \longrightarrow B + H_2O \qquad (v)$$

$$BH^+Cl^-(CH_3OH) \longrightarrow B(CH_3OH)H^+ + Cl^- \qquad (vi)$$

$$B(CH_3OH) \longrightarrow BH^+ + CH_3O^- \qquad (vii)$$

The high intensity absorption of the species $BH^+Cl^-$ indicates the presence of a system of conjugated double bonds. With increasing acidity of the solvent, the absorption due to $BH^+Cl^-$ diminishes and disappears

Table 4. *Ultraviolet absorption of β-phenethylbiguanide (602)*

| Compound | Solvent | $\lambda_{max}$ (mμ) | $\varepsilon \times 10^{-3}$ |
|---|---|---|---|
| β-Phenethylbiguanide HCl | Water | 233 | 14.5 |
| | $10^{-3}N$ HCl | 233 | 11.2 |
| | $10^{-2}N$ HCl | Non-specific absorption | |
| | $10^{-4}N$ NaOH | 233 | 14.5 |
| | $10^{-1}N$ NaOH | 232 | 12.7 |
| | $N$ NaOH | 225—228 (plateau) | 12.0 |
| | Methanol | 234 | 17.7 |
| β-Phenethylbiguanide 2 HCl | Water | 233 | 14.3 |
| Phenylbiguanide HCl | Water | 242 | 14.6 |
| | $5 \times 10^{-3}N$ HCl | 242 | 11.8 |
| | $10^{-2}N$ NaOH | 223—230 (plateau) | 12.4 |

at pH 2. In water, the dihydrochloride shows the same spectrum as $BH^+Cl^-$, indicating rapid hydrolysis of the doubly charged to the singly charged cation (equation iii). The spectrum in dilute alkali is identical with that in water, indicating the presence of the singly charged cation (equation iv). Increasing basicity of the solvent removes the proton from $BH^+$ and the resultant spectrum is that of β-phenethylbiguanide itself (equation v). In methanol, a hyperchromic shift is observed. These data and certain chemical properties (see Section VII F) suggest that β-phenethylbiguanide in methanol is transformed as shown by equation (vii).

The experimental results indicate that forms such as LXVI contribute significantly to the structure of $BH^+$. Other cyclic forms, open-chain forms (both conjugated and unconjugated) and a variety of ionic forms may also be written.

The electron-releasing character of the β-phenethyl-group should make its nitrogen the most susceptible to proton attack in (LXVI a). Again, the RN-group in (LXVI c) is the most basic site for attachment of a hydrogen bond. Since highly hypoglycemic biguanides exist which contain an RNMe- instead of a RNH-group (the structure of which cannot be represented by LXVI c), LXVI a is likely to be the preferred form *(602)*.

The disappearance of all but the non-specific absorption from the spectrum with increasing acidity reflects a loss of conjugation as the doubly charged cation is formed (equation ii): this transformation is represented by the formation of (LXVII)($BH_2^{++}$) which permits the widest charge separation.

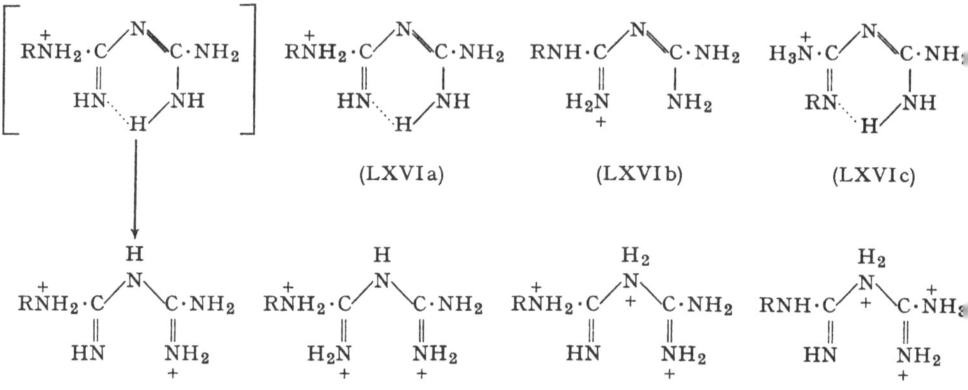

(LXVIa) (LXVIb) (LXVIc)

(LXVII) (LXVIIIa) (LXVIIIb) (LXVIIIc)

Since the molecule is stable as $BH_2^{++}Cl_2^{--}$, a third proton is required for hydrolysis, once the two potential sites for proton acceptance have been protonated. Addition of this proton is formulated in terms of structures (LXVIII a—c).

The ultraviolet spectra of a number of substituted arylbiguanides have been recorded and the effects of various substituents in both the aryl and biguanide moiety examined (*608*) (see Table 5). Similarities in the absorption properties of arylbiguanides and acetanilides are noteworthy.

As a result of these measurements, a structure (LXIXa) has been suggested for the arylbiguanide cation containing an aryl-group (sterically hindered by R') or an alkyl group at $N^1$ (in addition to the $N^1$-aryl group), and the term "biguanide resonance" has been applied to this type of structure. "Acetanilide resonance", typified by (LXIXb), is regarded as a significant factor where R = hydrogen and the aryl group is not sterically hindered. The authors claim (*608*) that forms such as (LXIXc) should show greater bathochromic shifts and extinction coefficients than are in fact observed.

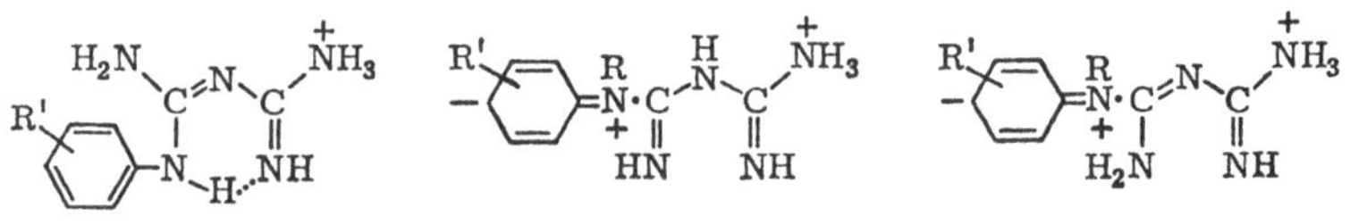

(LXIX a) (LXIX b) (LXIX c)

Table 5. *Ultraviolet absorption (608) of arylbiguanides*

$$R'-C_6H_4-NR\cdot C(=NH)\cdot NH\cdot C(=NH)\cdot NH_2\cdot HCl$$

| R′ | R | $\lambda_{max}$ mμ | $\varepsilon \times 10^{-3}$ |
|---|---|---|---|
| H | H | 242 | 14.6 |
| H | Me | 237 | 14.4 |
| H | Et | 236 | 16.5 |
| 2-Me | H | 236 | 15.6 |
| 2-Me | Et | 235 | 19.0 |
| 3-Me | H | 240 | 13.8 |
| 3-Me | Et | 236 | 17.4 |
| 4-Me | H | 240 | 15.0 |
| 2-Cl | H | 236 | 14.9 |
| 3-Cl | H | 249 | 13.7 |
| 4-Cl | H | 252 | 15.2 |
| 2-Br | H | 235 | 15.2 |
| 3-Br | H | 249 | 14.1 |
| 4-Br | H | 254 | 16.6 |
| 3-I | H | 228 | 25.0 |
| | | 243—261 | 14.2 |
| 3-$CF_3$ | H | 248 | 14.3 |
| 3-OH | H | 233—45 | 13.1 |
| | | 275—286 | 5.3 |
| 3-MeO | H | 235—251 | 13.4 |
| | | 274—287 | 4.6 |
| 4-$NH_2$ | H | 257 | 18.0 |
| 2,3-di-Me | H | 235 | 17.5 |
| 2,4-di-Me | H | 235 | 17.5 |
| 2,5-di-Me | H | 236 | 16.4 |
| 2,6-di-Me | H | 235 | 17.9 |
| 2,4,6-tri-Me | H | 235 | 20.4 |
| 2,6-di-Et | H | 235 | 18.2 |
| 2-Me-3-Cl | H | 234 | 16.9 |
| 2-Me-4-Cl | H | 236 | 19.3 |
| 2-Me-5-Cl | H | 237 | 14.4 |
| 2-Me-6-Cl | H | 236 | 16.8 |
| 2-Me-4-Br | H | 236 | 19.7 |
| 2,3-di-Cl | H | 232—244 | 14.1 |
| 3,5-di-Cl | H | 254 | 14.7 |
| 2-MeO-5-Cl | H | 287 | 4.7 |
| 4-$NH_2$-2,6-di-Cl | H | 243—252 | 13.1 |
| 2,5-di-MeO | H | 238—246 | 12.2 |
| | | 295 | 5.3 |

The ultraviolet spectra of a number of further biguanides, recorded during other investigations, are collected in Tables 6 and 7.

Table 6. *Ultraviolet absorption of biguanide derivatives*

| Compound | Solvent | $\lambda_{max}$ mμ | ε | Ref. |
|---|---|---|---|---|
| 1-p-Chlorophenylbiguanide | $10^{-2}N$ HCl | 250 | $1.245 \times 10^4$ | *(126)* |
| 1-β-Phenethylbiguanide HCl | Water | 234 | | *(166)* |
| 1-β-Phenethylbiguanide 2 HCl | Water | 234 | | *(166)* |
| 1-Benzyl-5-methylbiguanide HCl | Water | 236 | $1.65 \times 10^4$ | *(604)* |
| 1-Benzyl-5,5-dimethylbiguanide HCl | Water | 236 | $1.46 \times 10^4$ | *(604)* |
| 1-Benzyl-1,5-dimethylbiguanide HCl | Water | 238 | $1.55 \times 10^4$ | *(604)* |
| 1-Benzyl-1,5,5-trimethylbiguanide HCl | Water | 240 | $1.79 \times 10^4$ | *(604)* |
| 1,5-Dibenzylbiguanide HCl | Water | 237 | $1.78 \times 10^4$ | *(604)* |
| 1,5-Dibenzyl-2-methylbiguanide HCl | Water | 240 | $1.74 \times 10^4$ | *(604)* |
| 1,5-Dibenzyl-1,5-dimethylbiguanide HCl | Water | 244 | $1.89 \times 10^4$ | *(604)* |
| Paludrine | $10^{-2}N$ HCl | 250 | $1.18 \times 10^4$ | *(126)* |
| Paludrine | $10^{-1}N$ HCl | 236–246 (plateau) | $9.5 \times 10^3$ | *(653)* |
| Paludrine | Hexane | 264<br>230 | | *(73)* |
| Paludrine | HCl | 280 | | *(73)* |
| 1-(2-Methoxyphenyl)biguanide | Ethanol | 225<br>258<br>280 | $3.55 \times 10^4$<br>$3.16 \times 10^4$<br>$8.91 \times 10^3$ | *(74)* |
| 1-Phenylbiguanide | Ethanol | 259 | $2.0 \times 10^4$ | *(74)* |
| 1-Methyl-5-(4-methoxyphenyl)biguanide | Ethanol | 225<br>262<br>282 | $1.58 \times 10^4$<br>$5.01 \times 10^3$<br>$3.16 \times 10^3$ | *(74)* |
| 1-Isopropyl-5-(4-methoxyphenyl)-biguanide | Ethanol | 230<br>275 | $2.24 \times 10^4$<br>$5.63 \times 10^3$ | *(74)* |

Table 7. *Ultraviolet absorption of biguanide derivatives* (*78*)

$$NH_2\cdot\underset{\underset{NH}{\parallel}}{C}\cdot NH\cdot\underset{\underset{NH}{\parallel}}{C}\cdot N\langle_{H\quad R_3}^{R_1\ R_2}\rangle N\cdot\underset{\underset{NH}{\parallel}}{C}\cdot NH\cdot\underset{\underset{NH}{\parallel}}{C}\cdot NH_2\cdot 2HCl$$

| $R_1$ | $R_2$ | $R_3$ | Solvent | $\lambda_{max}$ mμ | $\varepsilon_{max} \times 10^{-4}$ |
|---|---|---|---|---|---|
| H | H | H | Water | 238—239 | 4.2 |
| Me | H | H | Water | 239—240 | 3.7 |
| Me | H | Me-trans | Water | 240—241 | 4.2 |
| Me | H | Me-cis | Water | 239 | 3.2 |
| Me | Me | H-trans | Water | 240 | 3.4 |
| Me | Me | H-cis | Water | 238—239 | 3.1 |

### *Validity of Beer's Law*

In the colorimetric estimation of β-phenethylbiguanide the plot of readings against concentration is found (*610*) to be a straight line at concentrations less than $10^{-4}M$. Above this concentration, β-phenethylbiguanide in solution deviates sharply from Beer's Law. It may be concluded that in the low concentration range over which Beers' Law is valid, the compound exists in the monomeric state in aqueous solution.

## 4. Infrared Spectra

Infrared data are available for biguanide (*688*) and 1,1-anhydrobis(β-hydroxyethyl)biguanide hydrochloride (*432*).

## 5. Nuclear Magnetic Resonance Spectra

An examination of the n.m.r. spectra of a number of alkylbiguanides has recently provided valuable evidence concerning the site of protonation in the biguanide-structure (*742a*). The measurements suggest a symmetrical structure (I l) for the monocation. Its further protonation occurs selectively and reversibly at N-3 to produce the dication (I m). Further protonation in concentrated sulphuric acid involves an attack of the proton on the π-electron system of (I m) at N-1, N-2, N-4 and N-5. The trication has consequently one of several possible prototropic structures, one of which is shown (I o).

## 6. Conductance

Conductance measurements of biguanide and its nitrate at various concentrations (*672*) have given values for their conductance ratio of the order of those of $\frac{1}{2}$ $Ba(OH)_2$, showing that biguanide is an exceptionally strong organic base. The second basic dissociation constant, determined by this method (*672*), is $2.24 \times 10^{-10}$.

Table 8. *Dissociation of biguanide: $B(OH)_2 \rightleftarrows BOH^+ + OH^-$*

| Concentration (moles/litre) | 0.1 | 0.05 | 0.01 | 0.004 | 0.001 |
|---|---|---|---|---|---|
| $\Lambda/\Lambda_\infty$ ($\Lambda_\infty = 234$) | 0.626 | 0.722 | 0.872 | 0.928 | 0.959 |

### 7. Nature of Ionic Species

The percentages of ionic species of 1-(8′-hydroxyquinol-5′-yl)biguanide ("5-biguanidoxine") present at pH 7.3 in water at 20° are (*9*): anion $<0.01\%$; cation, 97.9%; zwitterion, 2.1%; neutral molecule, none.

## C. Surface Tension

Values of the surface tension of solutions of Paludrine as a function of concentration have been recorded (*202*).

## D. Chromatography of Biguanides

Biguanide is absorbed by cation exchange resins and is eluted therefrom with hydrochloric acid. 1,1-Dimethylbiguanide, absorbed on Amberlite CG-120 resin, may be eluted (*508*) with solutions of sodium and potassium dihydrogen phosphate.

Elution curves for biguanide in the presence of related compounds (e.g. cyanamide, cyanoguanidine, guanidine, urea, thiourea) have been established (*690*). Various ion-exchange resins are suitable (*108, 686, 689*).

Data are available for the behaviour, during gas chromatography, of 1,1,2,2,4,4,5-heptamethyl and 1,1,2,2,5,5-hexamethylbiguanide (*110*).

# VII. Chemical Properties

In the following Sections the chemical properties of biguanide and its derivatives are systematically considered. Attention may be drawn to the particularly extensive and detailed recent studies concerning biguanides as sources of guanamines and pyrimidines (Section H). The basic nature and properties of biguanides have already been dealt with in Section VI B.

## A. Thermal Decomposition

### 1. Arylbiguanides

The thermal decomposition of phenylbiguanide, both in the presence and absence of ammonium chloride and aniline hydrochloride, has been studied in some detail (*669*) with the following results (equations 1 and 2, R = Ph):

$$\underset{\substack{\| \\ NH}}{RNH\cdot C\cdot NH_2} + H_2N\cdot CN \leftarrow RNH\cdot \underset{\substack{\| \\ NH}}{C}\cdot NH\cdot \underset{\substack{\| \\ NH}}{C}\cdot NH_2 \rightarrow H_2N\cdot \underset{\substack{\| \\ NH}}{C}\cdot NH_2 + RNH\cdot CN \quad (1)$$

$$PhNH\cdot CN + PhNH_2\cdot HCl \rightarrow PhNH\cdot \underset{\substack{\| \\ NH}}{C}\cdot NHPh\cdot HCl \quad (2)$$

In the presence of aniline hydrochloride, 1,3-diphenylguanidine hydrochloride is an additional product; its formation can be accounted for in four ways, but proceeds in fact exclusively according to equation (2) (*669*).

An additional by-product, m.p. 201—205° (27% relative to phenylbiguanide) was formulated (*669*) as either 2,4-diphenylmelamine or 1,2-diphenylisomelamine. The melting points (216—218° and 254°, respectively) of these two melamines subsequently synthesised (*375—377*) clearly suggest that the by-product was 2,4-diphenylmelamine.

### 2. Alkylbiguanides

The thermal decomposition of methylbiguanide hydrochloride proceeds similarly (*673*) (equation 1, R = Me). At the high temperatures, the cyanamide and methyl cyanamide formed trimerise to melamine derivatives. In the presence of an equimolar amount of ammonium chloride, biguanide and guanidine are obtained. Pyrolysis in the presence of one mole of methylamine hydrochloride resembles that of the aryl analogue in giving methylguanidine and guanidine, but differs from it in that the production of 1,3-dimethylguanidine was not confirmed.

According to the patent literature (*638*) pyrolysis of biguanide in closed vessels provides melamine in almost quantitative yields.

### 3. Acylbiguanides

A number of examples are recorded in the patent literature of the formation of guanamines (LXXI) by heating acyl (usually lauroyl or stearoyl) biguanides (LXX) to 170—220° (*53, 252, 267—272*).

$$RCO\cdot NH\cdot \underset{\underset{NH}{\|}}{C}\cdot NH\cdot \underset{\underset{NH}{\|}}{C}\cdot NH_2 \xrightarrow{170-220^\circ} \text{2,4-diamino-6-R-1,3,5-triazine } (H_2N,\ NH_2,\ R)$$

(LXX) (LXXI)

## B. Action of Alkalis

Arylbiguanides are unaffected (*608*) on being boiled in equimolar amounts of alkalis for short periods (30 min.). Alkylbiguanides are stable to alkalis at ambient temperatures (*602*), but both alkyl (*602*) and aryl (*126*) biguanides are decomposed at higher temperatures. Alkylbiguanides yield guanidines, ureas and amines after relatively short periods (2 hr.), but prolonged alkaline hydrolysis (18 hr. under nitrogen) cleaves arylbiguanides completely into amines and ammonia. Paludrine (*45*) gives p-chloroaniline after one hour's treatment with boiling alkali.

## C. Action of Mineral Acids

The acid hydrolysis of biguanides under a variety of conditions yields chiefly amidinoureas; the subject has been reviewed by *Ray* (*531*). The reaction proceeds slowly under very mild conditions at room temperature: thus, Paludrine, when stored in 2N hydrochloric acid slowly deposits N-p-chlorophenyl-N'-isopropylamidinourea hydrochloride (*140*). That this product was not the isomeric N-isopropyl-N'-p-chlorophenylamidinourea was verified by its unequivocal synthesis from isopropylguanidine and p-chlorophenyl isocyanate (*140*):

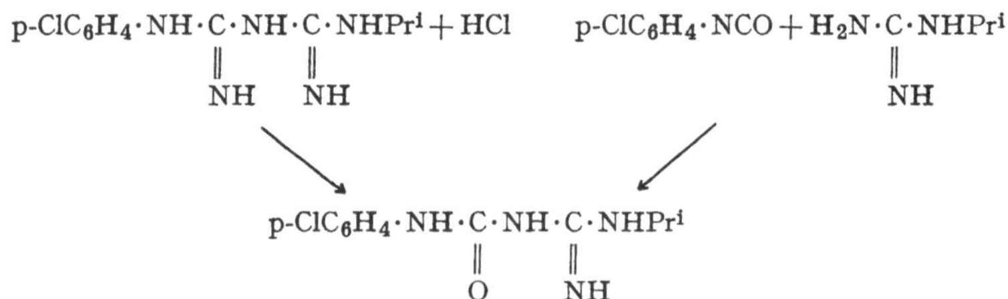

The mild acid hydrolysis of p-(substituted)sulphonamidophenylbiguanides to the corresponding amidinoureas proceeds rapidly and almost quantitatively (*368*). The presence of a phenyl- (*368*) or alkyl- (*370*) group in position $N^5$ of these sulphobiguanides inhibits hydrolysis, but comparable substituents in the p-sulphonamidophenyl-group are without effect (*370*). A kinetic study (*369*) of the effect of substituents in the sulphonamido-group, and in position $N^5$ of the biguanide molecule has shown that the hydrolysis is usually, a unimolecular reaction. Velocity constants, at 32° and 46°, for eight compounds are available (*369*).

Conversion to the amidinourea occurs rapidly (20 min.) under the influence of boiling acids (*383*, *439*); heterocyclic biguanides are reported (*589*) to give amidinoureas in 60—80% yield after 30—45 min. refluxing in 12% hydrochloric acid. However, on more prolonged action, the biguanide chain may be cleaved: m-chlorophenylbiguanide, for example, is decomposed into N-amidino-N'-m-chlorophenylurea (28%) and m-chloroaniline (47%) (*608*).

In view of the generally easy hydrolysis of biguanides, it is surprising to find (*602*) that β-phenethylbiguanide is resistant to the action of acids. After prolonged heating with 3*N* and 6*N* hydrochloric acid, 50% sulphuric acid or polyphosphoric acid, the diacid salt is recovered in each case. Possible explanations of this observation have been discussed (*602*).

A detailed quantitative study of the acid hydrolysis of biguanides has been carried out (*653*) using Paludrine and three related compounds (LXXII—LXXV) under carefully controlled conditions: the results clearly showed that the rate of hydrolysis decreases with increasing degree of substitution in the biguanide-structure.

$p\text{-}ClC_6H_4\cdot NH\cdot C(=NH)\cdot NH\cdot C(=NH)\cdot NH_2$ p-Chlorophenylbiguanide (LXXII)

$p\text{-}ClC_6H_4\cdot NH\cdot C(=NH)\cdot NH\cdot C(=NH)\cdot NHPr^i$ Paludrine (LXXIII)

$p\text{-}ClC_6H_4\cdot NH\cdot C(=NH)\cdot NH\cdot C(=NH)\cdot NMePr^i$ 1-p-Chlorophenyl-5-methyl-5-isopropylbiguanide (LXXIV)

$p\text{-}ClC_6H_4\cdot NH\cdot C(=NH)\cdot NH\text{-}$(pyrimidin-2-yl, 4-$CH_3$, 6-$NH(CH_2)_2NEt_2$) 2-p-Chlorophenylguanidino-4-(ω-diethylamino)ethylamino-6-methylpyrimidine (LXXV)

Both Paludrine and its 5-methyl-homologue (LXXIV) are quantitatively hydrolysed on being heated with 0.25 *N* hydrochloric acid under pressure (20—25 p.s.i.) during twelve hours (*652*). This quantitative reaction is of practical importance: the resulting p-chloroaniline may be diazotized and coupled to give a dye, thus providing a basis for the estimation of Paludrine.

The synthesis of biguanides by condensation of arylamines and cyanoguanidine (Section III B) requires acid conditions. In the presence of too high a concentration of acid, however, amidinoureas will result (*572, 730*). Thus, the appropriate reactants give the desired 1-arylbiguanides in boiling 12% hydrochloric acid, but yield N-amidino-N'-arylureas in 22% acid. Since 22% hydrochloric acid converts the biguanide into the amidinourea (*730*), support is provided for the primary formation of the biguanides.

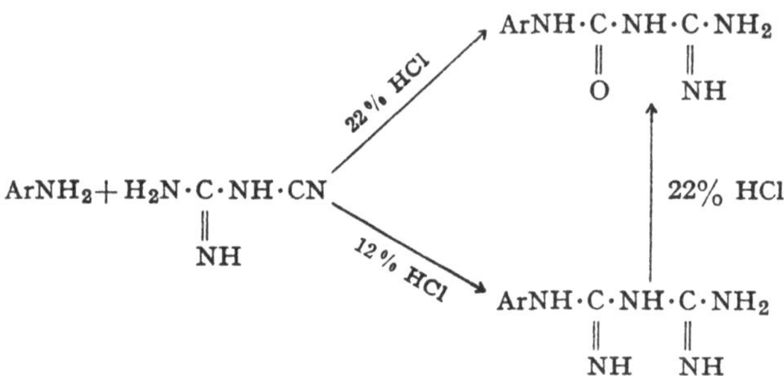


Time, as well as acid concentration, is a factor controlling the nature of the product. Thus p-bromoaniline and cyanoguanidine successively yield the corresponding biguanide and amidinourea, either of which may be isolated at the appropriate time (*729*).

## D. Action of Nitrous Acid

The interaction of biguanide and nitrous acid yields cyanoguanidine. *Pellizzari* (*497*), who first described this reaction, considered that it in-

$$H_2N{\cdot}\underset{\underset{NH}{\|}}{C}{\cdot}NH{\cdot}\underset{\underset{NH}{\|}}{C}{\cdot}NH_2 + HNO_2 \longrightarrow \left[H_2N{\cdot}\underset{\underset{NH}{\|}}{C}{\cdot}NH{\cdot}\underset{\underset{NH}{\|}}{C}{\cdot}NH{\cdot}NO\right]$$

(LXXVI)

$$\rightarrow H_2N{\cdot}\underset{\underset{NH}{\|}}{C}{\cdot}NHCN + N_2 + H_2O$$

volved the formation of an intermediate of type LXXVI [the reaction path resembling that of the corresponding reaction of guanidine (*694*)], and not the direct withdrawal of a molecule of ammonia as ammonium nitrite which would yield the by-products on further decomposition. During studies of the Van Slyke determination of biguanide, *Rosenthaler* (*552*) showed that less than half a mole of nitrogen is evolved per mole of biguanide, but the other reaction products were not examined.

In contrast, 1-substituted biguanides are hydrolysed to N-amidino-N'-substituted ureas (LXXVII) (*498*), the structure of which is confirmed by the unequivocal synthesis of their isomers (LXXVIII) by acid hydrolysis of the appropriate cyanoguanidine.

$$\text{RNH}\cdot\underset{\text{NH}}{\underset{\|}{\text{C}}}\cdot\text{NH}\cdot\underset{\text{NH}}{\underset{\|}{\text{C}}}\cdot\text{NH}_2 \rightarrow \text{RNH}\cdot\underset{\text{O}}{\underset{\|}{\text{C}}}\cdot\text{NH}\cdot\underset{\text{NH}}{\underset{\|}{\text{C}}}\cdot\text{NH}_2 \quad \text{(LXXVII)}$$

$$\text{RNH}\cdot\underset{\text{NH}}{\underset{\|}{\text{C}}}\cdot\text{NH}\cdot\underset{\text{O}}{\underset{\|}{\text{C}}}\cdot\text{NH}_2 \leftarrow \text{RNH}\cdot\underset{\text{NH}}{\underset{\|}{\text{C}}}\cdot\text{NHCN}$$

(LXXVIII)

Nitrous acid converts 1-aryl-5-alkylbiguanides into N-aryl-N'-alkylamidinoureas, and *not* N-alkyl-N'-arylamidinoureas, as shown by the unequivocal synthesis of the former (*140*). Alkylbiguanides behave (*602*) as the foregoing aromatic analogues, excepting β-phenethylbiguanide which did not give pure products (*602*).

$$\text{PhNH}\cdot\underset{\text{O}}{\underset{\|}{\text{C}}}\cdot\text{NH}\cdot\underset{\text{NH}}{\underset{\|}{\text{C}}}\cdot\text{NHPh} \longleftarrow \text{PhNH}\cdot\underset{\text{NH}}{\underset{\|}{\text{C}}}\cdot\text{NH}\cdot\underset{\text{NH}}{\underset{\|}{\text{C}}}\cdot\text{NHPh} \not\longrightarrow \text{(LXXX)}$$

(LXXIX)

(LXXX): 4,6-dianilino-*as*-tetrazine ring, PhNH and NHPh substituents

Wystrach (*757*) attempted to employ this reaction with a different object: 1,5-Diphenylbiguanide was treated with *two* moles of nitrous acid at 25°, with the aim of obtaining the unknown 4,6-dianilino-*as*-tetrazine (LXXX). However, the products were in fact the amidinourea (LXXIX) and one mole of nitrogen per mole of diphenylbiguanide. *Wystrach* has therefore suggested that the low recovery of nitrogen observed by *Rosenthaler* (*552*) may indicate the possible formation of some *as*-tetrazine in this reaction; its careful reinvestigation might be rewarding.

## E. Oxidation and Reduction

### 1. Action of Oxidising Agents

Although biguanides are reactive compounds, the biguanide chain appears to be resistant to oxidation. The parent compound, biguanide, does not react with iodine (*372*); Paludrine gave no recognisable products when treated with lead tetra-acetate, potassium ferricyanide, hydrogen peroxide and acid or alkaline potassium permanganate (*56*).

### 2. Action of Reducing Agents

In general, the biguanide structure appears to be fairly stable towards reducing agents. Iron and mineral acids, for example, appear to be without effect, several cases (*521*, *548*, *753*) being known in which 1-(p-nitrobenzenesulphonyl)biguanides are reduced to the corresponding amino-compounds by this reagent. Similarly, the biguanide-moeity remains intact when biguanides are debenzylated by palladium on carbon (*416*, *604*).

The action of hypophosphorous acid, both in the presence (*556*) and absence (*408*) of potassium iodide, on p-biguanidophenylarsonic acid gives p,p′-bis-biguanidoarseno-benzene, the biguanide portion being again unaffected:

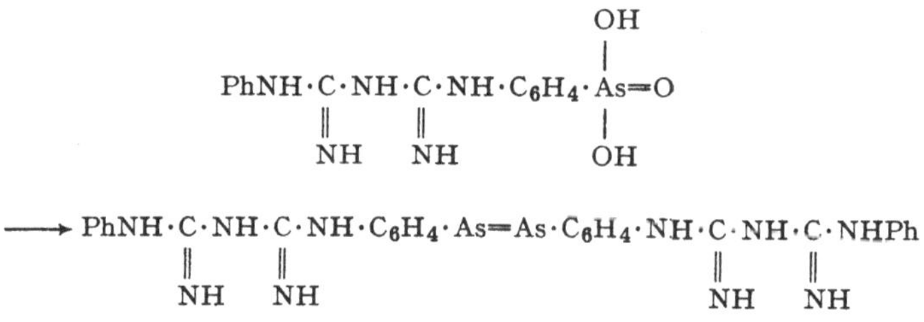

$$PhNH\cdot \underset{\displaystyle NH}{\overset{\displaystyle \|}{C}}\cdot NH\cdot \underset{\displaystyle NH}{\overset{\displaystyle \|}{C}}\cdot NH\cdot C_6H_4\cdot \overset{\displaystyle OH}{\underset{\displaystyle OH}{As}}{=}O$$

$$\longrightarrow PhNH\cdot \underset{\displaystyle NH}{\overset{\displaystyle \|}{C}}\cdot NH\cdot \underset{\displaystyle NH}{\overset{\displaystyle \|}{C}}\cdot NH\cdot C_6H_4\cdot As{=}As\cdot C_6H_4\cdot NH\cdot \underset{\displaystyle NH}{\overset{\displaystyle \|}{C}}\cdot NH\cdot \underset{\displaystyle NH}{\overset{\displaystyle \|}{C}}\cdot NHPh$$

Paludrine is cleaved by zinc amalgam and acid to a primary amine (*615*), which may be diazotised *in situ* and coupled to yield a dye; a colorimetric estimation of Paludrine is based on this reaction sequence.

## F. Alkylation of Biguanides

The alkylation of acyl (usually lauroyl or stearoyl) biguanides, generally by means of dimethyl sulphate at 100—110°, has been the subject of numerous patents (*53*, *54*, *234*, *241—245*, *248—251*, *275*, *277*). However, the structure of the final products was not specified except in one case (*275*) where the position of the methyl-group remained again undefined ("methyl 2-stearoyl-5-phenylbiguanide").

Another example concerns the action of isopropyl iodide in isopropanol on the "stable form" of Paludrine (*45*) (see Section III N), an unfortunate choice of starting material. The product is stated to be 1-p-chlorophenyl-2-isopropylbiguanide hydriodide [p-$ClC_6H_4NH \cdot C(:NPr^i) \cdot NHC(:NH)$-$NH_2 \cdot HI$].

β-Phenethylbiguanide is reported (*602*) to be resistant to methylation but as it behaves generally somewhat anomalously, e.g. in its stability towards acids (Section VII C), this provides no information as to the general behaviour of biguanides on alkylation. Thus, methyl iodide in methanol converted β-phenethylbiguanide merely into its hydriodide, and similar results were observed with methyl tosylate, and benzyl bromide.

It is these reactions that are thought to provide some justification for equation (vii) [in Section VI B3] according to which β-phenethylbiguanide in methanol exists to some extent as biguanidium methoxide. This being so, the reaction with alkyl halides is the nucleophilic displacement of the halogen ion by methoxide with formation of the methyl ether and the salt of the biguanide.

## G. Acylation of Biguanides

The conversion of biguanides into guanamines under the influence of acylating agents is well known (see also Section VII I2). Acetic anhydride or benzoyl chloride in conjunction with alkali react with the parent compound, yielding 2-methyl- (*519*) or 2-phenyl-4,6-diamino-s-triazine (*17*) (LXXXI; R = Me or Ph) respectively.

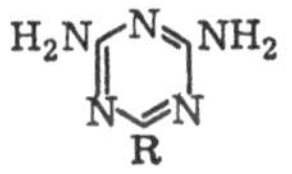

$H_2N$, N, $NH_2$ / N, N / R (LXXXI) RNH, N, NHR' / N, N / $CH_3$ (LXXXII)

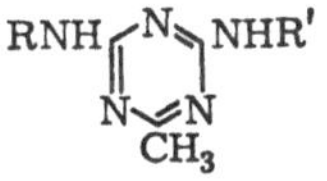

The acetylation of p-chlorophenylbiguanide has accordingly, by analogy, been formulated (*147*) to yield 2-methyl-6-p-chlorophenylguanamine (LXXXII; R = p-$ClC_6H_4$, R' = H), and that of Paludrine (*217*) to yield the isopropyl analogue (LXXXII; R = p-$ClC_6H_4$, R' = $Pr^i$).

### *Arylsulphonyl Halides*

In contrast, sulphonyl halides acylate biguanides in the expected manner, the parent compound, for example, being convertible into 1-*p*-nitrobenzenesulphonylbiguanide. The reaction has been extended to 1,1-disubstituted biguanides (*548*) and to several other examples (*167, 521*).

## H. Condensation with Esters

The reaction of biguanides with esters was first studied in detail by *Rackmann* (*518*) who showed that ethyl formate and arylbiguanides gave 2,4-diamino-6-substituted s-triazines (LXXXIII, where R′, R″ and R‴ may vary widely).

$$\underset{\qquad\quad \|\qquad\ \|\atop \qquad\quad NH\qquad NH}{R'NH\cdot C\cdot NH\cdot C\cdot NHR''} + R'''\cdot COOAlk \longrightarrow \text{(LXXXIII)} + AlkOH + H_2O$$

(LXXXIII): triazine ring bearing R‴, NHR″ and NHR′

### 1. Saturated Esters

In 1940, phenylbiguanide and ethyl formate were reported (*735*) to react at room temperature to give a triazine, also formed on fusion of phenylbiguanide and diphenylformamide at 145°. Further experience gained with Paludrine (*155*) showed that the products were $N^2$-arylguanamines (LXXXIII; R′ = Ph, R″ = R‴ = H; *or* R′ = p-$ClC_6H_4$; R″ = $Pr^i$; R‴ = H). The condensation thus involves the $N^2$ and $N^4$ positions of the biguanide molecule. The cyclisation is widely applicable to 1-substituted biguanides (*607*) and generally proceeds in good yields. The arylbiguanide salt of formic acid is isolated as a by-product.

#### a) *Biguanide*

During the 1940's the chemistry of guanamines (LXXXIII) attracted much attention, especially from investigators of the American Cyanamid Company who reported in the patent literature their observations on the interaction of biguanides and a wide range of carboxylic acid esters.

Formoguanamine (LXXXIII; R′ = R″ = R‴ = H) was readily accessible (*697*) from biguanide and formates, but yields diminished sharply when esters of higher carboxylic acids were employed. However, metal alkoxides proved excellent condensing agents which promoted the formation of guanamines containing long aliphatic chains (e.g. 2-stearoguanamine, 2-lauroguanamine, etc.) in good yields. This catalytic effect of the alkoxides appears to be primarily chemical, as increasing amounts (optimum: equivalent quantities) raise both the speed and yield of the reaction. Various metal alkoxides are suitable, and methanol and ethanol are the preferred solvents.

The condensation of biguanide and methyl dodecanoate at room temperature, for example, affords 83% yields of (LXXXIII; R′ = R″ = H, R‴ = $C_{11}H_{23}$) in the presence of sodium ethoxide (1 mole), but only 54% yields in its absence after 18 hours' refluxing. The use of ethyl difluoroacetate ($F_2CHCOOEt$) and related esters in this reaction has also been reported (*365*).

b) *Alkyl- and Arylbiguanides*

1-Alkyl or 1-arylbiguanides (*697*) react analogously: 1-Arylbiguanides condense slowly with esters other than those of formic acid (*226*), but metal alkoxide-catalysts (*475*) or caustic alkalis (*477*) have again been employed (*475*) to good effect. The synthesis has been widely used (*19, 127, 204, 221, 417, 593, 599, 601, 602, 677, 701*).

Esters of aromatic (instead of aliphatic) acids react rapidly at room temperature (*701*). but the addition of a condensing agent is still desirable when 1-substituted biguanides are used.

Heterocyclic carboxylic esters (*708*) are as reactive as formate and acetate esters. Extended to 1-substituted biguanides, the reaction is promoted by metal alkoxides. Since the heterocyclic ring must be five- or six-membered and contains a conjugated system of double bonds, the reaction resembles formally that involving α,β-unsaturated esters (see below). Several successful examples are on record (*599, 606*).

2. α,β-Unsaturated Esters

a) *Biguanide*

α,β-Unsaturated esters (*703*) are particularly active, participating in smooth and rapid condensations with the parent base, even in the absence of catalysts, although these are desirable for 1-substituted biguanides. A negative substituent on the α-carbon atom of the ester exerts yet additional activation.

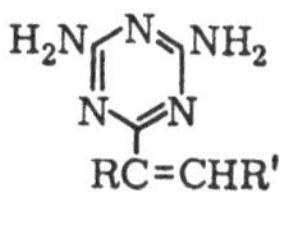

| H₂N-triazine-NH₂ with RC=CHR' | H₂N-triazine-NH₂ with CH₂CH(OR)CH₃ | H₂N-triazine-NH₂ with CH₂CH₂OMe |
|---|---|---|
| (LXXXIV) | (LXXXV) | (LXXXVI) |

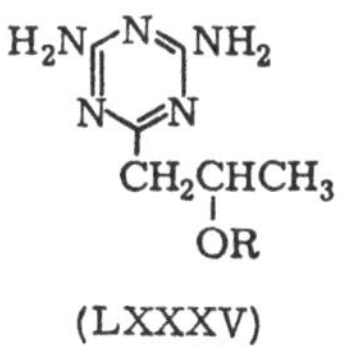

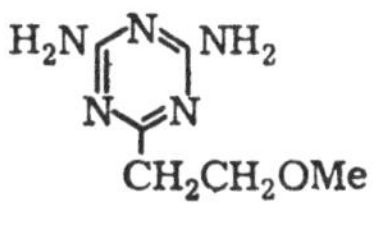

The double bond of certain α,β-unsaturated esters (e.g. crotonates, methacrylates) is so reactive that it tends to add the elements of the alcohol present as solvent; the resulting β-alkoxy-substituted guanamines (LXXXV) (*79*) are therefore readily and cheaply produced, and the use of expensive β-alkoxycarboxylic esters required in the conventional synthesis (*706*) is avoided. Crotonic esters give the alkoxyguanamine

(LXXXV) exclusively; methacrylic esters afford the α,β-unsaturated guanamine (LXXXIV; R = Me; R′ = H) as the main product (61%), the alkoxyguanamine appearing as a by-product.

The condensation of biguanide with acrylate esters [e.g. methyl acrylate in presence of sodium methoxide (*484*)] was expected to afford a vinyl-triazine (LXXXIV, R = R′ = H), but gave in fact 2-(β-methoxyethyl)guanamine (LXXXVI) (51%). The desired vinyl-triazine was finally synthesised (*486*) from phenylbiguanide and acrylyl chloride in aqueous acetonitrile and its structure proved by its hydrogenation to 2-ethyl-6-phenylguanamine.

b) *Arylbiguanides*

The reaction between phenylbiguanide and acrylate esters in the presence of molar quantities of sodium ethoxide has been studied in detail (*486*). The available evidence suggests that the condensation proceeds as follows:

$$CH_2{=}CH{\cdot}\overset{\overset{O}{\|}}{C}{\cdot}OR' + {}^{-}OR + H^+ \longrightarrow RO{\cdot}CH_2{\cdot}CH_2{\cdot}\overset{\overset{O}{\|}}{C}{\cdot}OR' \qquad \text{(i)}$$

$$RO{\cdot}CH_2{\cdot}CH_2{\cdot}\overset{\overset{O}{\|}}{C}{\cdot}OR' + PHhN{\cdot}\underset{\underset{NH}{\|}}{C}\overset{NH}{\diagup\diagdown}\underset{\underset{NH}{\|}}{C}{\cdot}NH_2 \longrightarrow \text{H}_2\text{N-triazine(NHPh)(CH}_2\text{CH}_2\text{OR)} \qquad \text{(ii)}$$

$$CH_2{=}CH{\cdot}\overset{\overset{O}{\|}}{C}{\cdot}OR' + {}^{-}OR \longrightarrow RO{\cdot}CH_2{\cdot}CH_2{\cdot}\overset{\overset{O}{\|}}{C}{\cdot}ONa + R'OH \qquad \text{(iii)}$$

The competing reaction (iii) cannot be entirely suppressed. In the absence of a catalyst, the same reaction is claimed (*486*) to give the open chain adduct [PhNH·C(:NH)NHC(:NH)·$NHCH_2CH_2COOH$].

Esters containing two double bonds display activation and are suitable reactants in the guanamine synthesis (*705*).

In one case, namely the interaction of phenylbiguanide and ethyl cinnamate, the product is reported (*642*) to be a pyrimidine (LXXXVII).

$$PhNH{\cdot}\underset{\underset{NH}{\|}}{C}{\cdot}NH{\cdot}\underset{\underset{NH}{\|}}{C}{\cdot}NH_2 + PhCH{=}CH{\cdot}CO_2Et \longrightarrow PhNH{\cdot}\underset{\underset{NH}{\|}}{C}{\cdot}NH\text{-pyrimidinone (4-Ph, HN, N, =O)}$$

(LXXXVII)

### 3. Esters containing Negative Groups

Esters activated by a negative group (e.g. cyano, halogen, carboxy, carbalkoxy, sulpho, nitro, acid amide, carbonyl, hydroxy, alkoxy, etc.) are particularly reactive in this general synthesis of guanamines (*699*). Surprisingly, this activation appears to be independent within wide limits of the position of the negative group in the chain, as is illustrated by the following two examples:

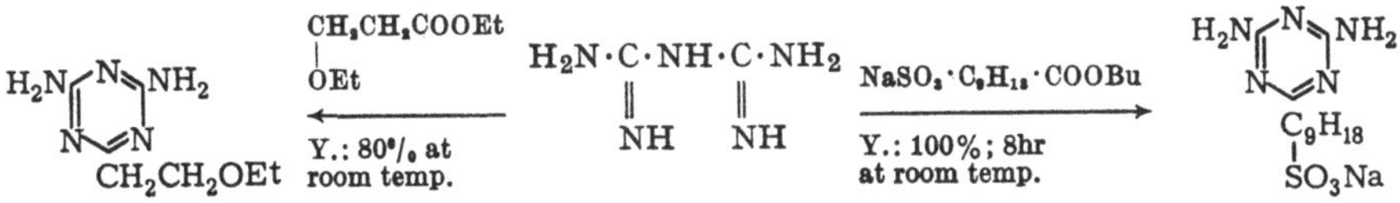

$H_2N\cdot C\cdot NH\cdot C\cdot NH_2$ (with =NH on each C) + $CH_2CH_2COOEt$ / OEt → (Y.: 80% at room temp.) 2,4-diamino-6-($CH_2CH_2OEt$)-triazine; + $NaSO_3\cdot C_9H_{18}\cdot COOBu$ → (Y.: 100%; 8hr at room temp.) 2,4-diamino-6-($C_9H_{18}SO_3Na$)-triazine

In reactions involving 1-substituted biguanides, the activating influence of negative groups is less effective, so that addition of a metal alkoxide may be necessary to ensure satisfactory yields of 2-substituted guanamines (*609, 699*). Disubstituted biguanides require vigorous conditions (*609*).

Numerous patents have dealt with condensations of this type, each treating one particular activating group in detail (e.g. halogen-esters (*704*), keto-esters (*696*), sulpho-esters (*698*)).

#### a) *Halogeno-Esters*

Haloesters have been particularly widely used (*484, 599, 600, 602, 605, 606, 609*) in this synthesis. Amongst the large number of examples studied, an anomalous reaction has been noted. While phenylbiguanide and mono- and di-chloroacetic esters afford the expected guanamines in good yields (54—61%) (*600*), ethyl trichloroacetate gives almost exclusively (84%) 2-amino-4-anilino-6-hydroxy-triazine (LXXXIX), the expected trichloro-substituted guanamine (XC) arising only in traces (1.3%). The course of this reaction has been formulated as follows (*600*):

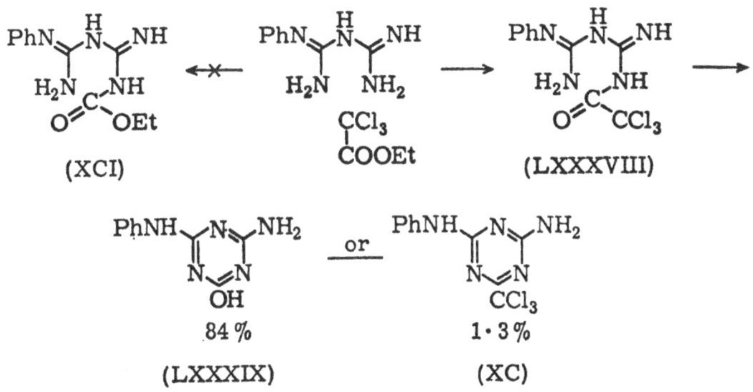

PhN=C(NH₂)–NH–C(=NH)–NH₂ + $CCl_3COOEt$ —✕→ (XCI) [$PhN$, $H_2N$, NH, NH, $C(=O)OEt$]; → (LXXXVIII) [$PhN$, $H_2N$, NH, NH, $C(=O)CCl_3$] →

(LXXXIX) 2-anilino-4-amino-6-OH-triazine, 84 % or (XC) 2-anilino-4-amino-6-$CCl_3$-triazine, 1·3 %

The facile elimination of the trichloromethyl anion favours the formation of (LXXXIX) and provides the driving force for triazine formation. The alternative possible initial formation of (XCI) seems less likely, since cyclohexylamine and ethyl trichloroacetate give cyclohexyltrichloroacetamide ($C_6H_{11}\cdot NH\cdot CO\cdot CCl_3$) and not the cyclohexylurethan (*600*).

The above results were verified in a later study (*605*), but several other halogenated esters ($R\cdot CO_2Et$; $R=CF_3$; $C_2F_5$; $C_3F_7$; $CFCl_2$; or $CHBr_2$) did not show this effect. The mechanism postulated (*605*) is essentially that given before (*600*); it is noteworthy however that the detailed structure of the biguanide employed as reactant in methanol differs from that given in an earlier paper (*602*) by the same authors, thus introducing some inconsistency into the discussion of the detailed mechanism.

The interaction of phenylbiguanide with ethyl α-bromopropionate (but *not* with the corresponding chloro-ester) in methanol-sodium methoxide (*600*) yields phenylbiguanide hydrobromide in addition to the expected 2-(α-bromoethyl)-6-phenylguanamine (19%). It seemed possible that this salt had arisen by dehydrohalogenation of the guanamine; the likelihood of this view was supported by the observation that interaction of the guanamine and phenylbiguanide gave phenylbiguanide hydrobromide and a polymeric residue. Phenylbiguanide and ethyl β-bromopropionate similarly gave phenylbiguanide hydrobromide and ethyl acrylate by dehydrohalogenation; the latter condensed with phenylbiguanide under the prevailing conditions to 2-(β-methoxyethyl)-6-phenylguanamine (i.e. the phenyl derivative of LXXXVI).

The reaction between 2-(α-bromoethyl)-6-phenylguanamine and phenylbiguanide was carried out (*486*) in an unsuccessful attempt to obtain a vinyl-guanamine; phenylbiguanide hydrobromide, however, was recovered.

### b) *Hydroxy-Esters*

An anomalous reaction of the hydroxy-activated ester, ethyl lactate has also been noted (*602*). With β-phenethylbiguanide this gave, in addition to the expected guanamine (XCII), the corresponding hydroxy-compound (XCIII), but no chemical evidence for this formulation was given.

$H_2N$–triazine–$CH(OH)CH_3$, $NHCH_2CH_2Ph$

(XCII)

$HO$–triazine–$CH(OH)CH_3$, $NHCH_2CH_2Ph$

(XCIII)

c) *Keto-Esters*

i) *Ethyl Acetoacetate.* In their condensation with biguanides, ethyl acetoacetate and other keto-esters (*150, 696*) react in two distinct ways; attack at $N^4$ and $N^5$ of the biguanide molecule yields pyrimidines, but interaction at positions $N^2$ and $N^4$ (or $N^5$) results in triazines. The condensation of keto-esters and the parent compound, biguanide, normally proceeds readily in the absence of condensing agents, but addition of a metal alkoxide is usually advantageous for substituted biguanides.

1-*Monosubstituted Biguanides.* 1-Monosubstituted biguanides can presumably undergo both types of condensation with equal ease, the prevailing reaction depending on the nature of the 1-substituent. In the interaction of p-chlorophenylbiguanide and ethyl acetoacetate (*147*) the main product (80%) was the appropriate pyrimidine (XCIV, R= =p-$ClC_6H_4$), a small amount of by-product being formulated as the triazine (XCV; R=p-$ClC_6H_4$). Similar parallel condensations have been described (*150, 298, 417*), including examples involving 1-heteryl-biguanides (*584*).

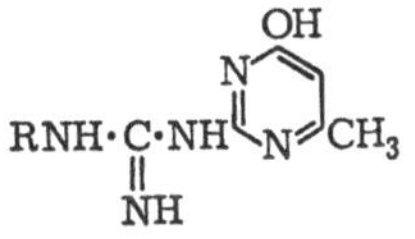

(XCIV) 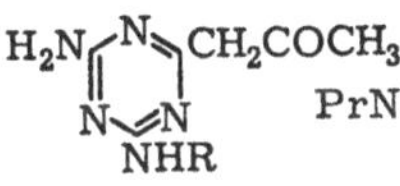 (XCV) 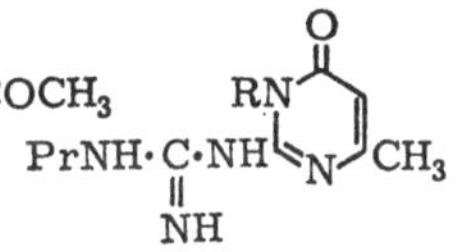 (XCVI) 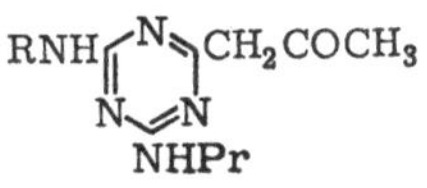 (XCVII)

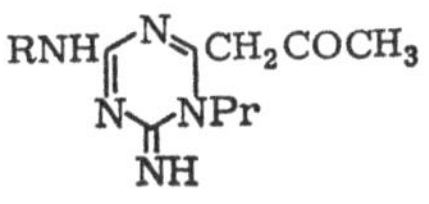

(XCVIII) 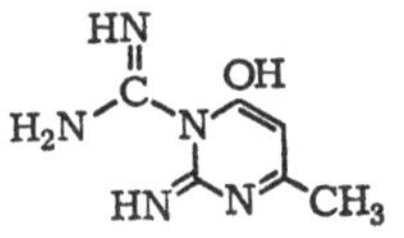 (XCIX) 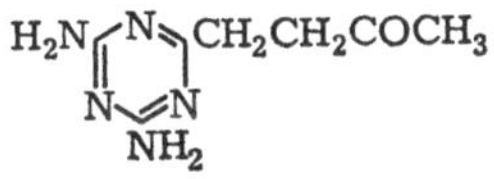 (C)

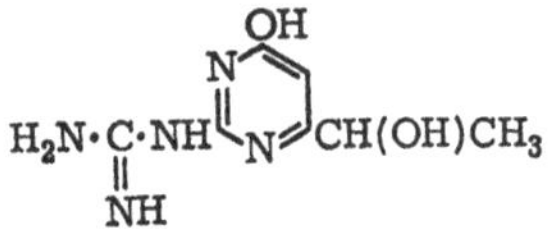

(CI) 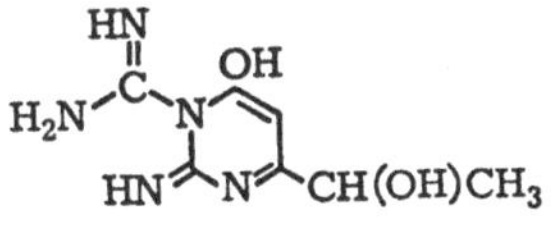 (CII) 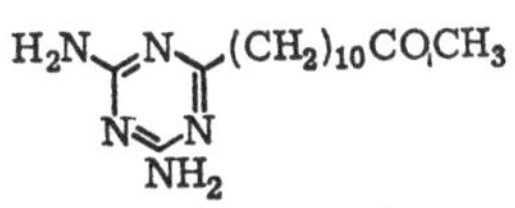 (CIII)

The interaction of 1-(p-arsonophenyl)biguanide and ethyl acetoacetate failed to give the expected 2-(p-arsonophenylguanidino)-4-hydroxy-6-methylpyrimidine under a variety of conditions (*408*), as did the corresponding reaction with diethyl malonate. Condensation did occur, however, with acetylacetone (see Section VII J 5) (see also ref. *590*).

1,5-*Disubstituted Biguanides.* Since the "blocked" $N^5$-position of 1,5-disubstituted biguanides is less likely to participate in condensations, the formation of pyrimidines might be expected to be inhibited. Paludrine was studied in detail (*217*) from this point of view, since a condensation of this type might occur in the living body, and its elucidation might contribute to our understanding of its drug mechanism.

Condensation under the usual conditions gave in fact the pyrimidine (XCVI, R = p-$ClC_6H_4$) in only 10—15% yield. Even an excess of ethyl acetoacetate raised the yield only inconsiderably (to 30%), the rest of the product being a brown viscous oil, from which the supposed triazine (XCVII) (arising by $N^2$,$N^4$-addition) was isolated. The alternative possible $N^2$,$N^5$-addition would result in the theoretically possible structure (XCVIII). However, as Paludrine is known to react with ethyl formate at $N^2$,$N^4$ (*372*), it is assumed by analogy that ethyl acetoacetate reacts similarly.

ii) *Other Keto-Esters.* A keto-group, whatever its position in an ester activates the condensation, but its location is important in determining whether or not a pyrimidine is formed. With β- and γ-keto-esters some pyrimidine is always produced.

Methyl acetoacetate is reported (*696*) to yield the triazine (XCV; R = H) as main product (85%), the pyrimidine being the by-product, of probable structure (XCIX) rather than (XCIV; R = H). However, in view of the opposite conclusion in the case of the condensation of ethyl acetoacetate and substituted biguanides (see above), this formulation requires careful confirmation.

The interaction of biguanide and ethyl 4-ketopentanoate affords the triazine (C) and pyrimidine (CII?) in equal yields (40% each) (*696*). The structure of the latter (CI or CII) was not confirmed, but on the basis of the results of the corresponding reaction of ethyl acetoacetate (see above), structure (CI) seems more likely.

The reaction of biguanide and ethyl luvinate also affords equal proportions of the expected triazine and the pyrimidine (*699*).

Esters containing a keto-group beyond the γ-position cannot give rise to pyrimidines, but are still highly activated, as is shown by the excellent yields (93%) of (CIII) obtained from biguanide and 2-keto-tridecanoate (*699*).

### d) *Cyano-Esters*

Cyano-esters react with biguanides with results not unlike those of keto-esters. Thus, biguanide and methyl 2-cyano-4-ethyloct-3-enoate (*699, 703*) [$CH_3(CH_2)_3CHEt \cdot CH{=}CH \cdot CH(CN)CO_2Me$] give a product which may be either a triazine or a pyrimidine, but a choice has so far not been made. Similar doubt exists regarding the product from biguanide and ethyl cyanoacetate: it was formulated as a pyrimidine (*10*), but is stated to be a triazine in the patent literature (*699*).

### 4. Lactones

Lactones, which may be regarded as internal esters, react with biguanides yielding guanamines (*476*) (e.g. CIV). Their reactivity varies with their structure; some lactones require the presence of metal alkoxides as condensing agents, and these are also desirable when substituted biguanides are used. In the non-catalysed condensation of phenylbiguanide and propiolactone, the isolation of an open-chain compound (CVI) in addition to the expected triazine (CV) has been claimed (*486*). The structure of (CVI) was assigned on the basis of its analogy with the product obtained from the non-catalysed reaction of phenylbiguanide and ethyl acrylate.

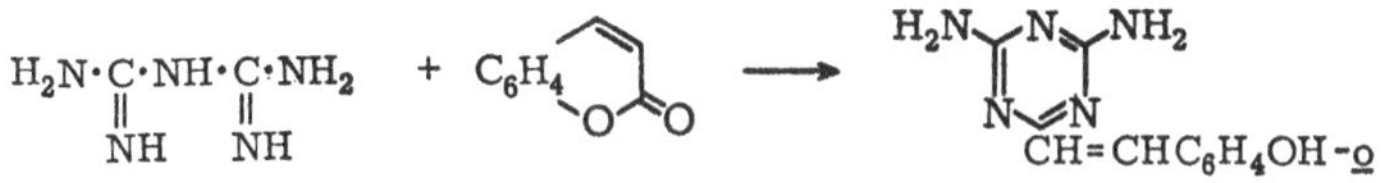

$H_2N \cdot C(=NH) \cdot NH \cdot C(=NH) \cdot NH_2$ + $C_6H_4$ (lactone, O, O) ⟶ $H_2N$, N, $NH_2$ / N, N / $CH{=}CHC_6H_4OH$-o

(CIV)

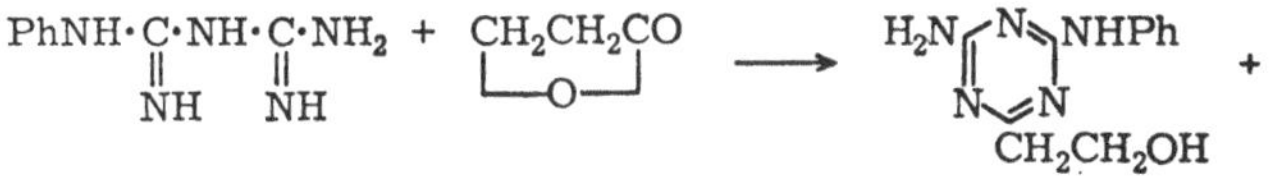

$PhNH \cdot C(=NH) \cdot NH \cdot C(=NH) \cdot NH_2$ + $CH_2CH_2CO$ (—O—) ⟶ $H_2N$, N, NHPh / N, N / $CH_2CH_2OH$

(CV) 18%

+

$PhNH \cdot C(=NH) \cdot NH \cdot C(=NH) \cdot NH \cdot COCH_2CH_2OH$

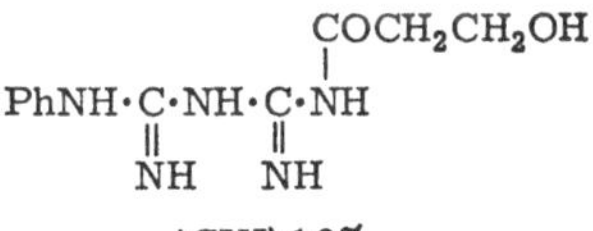

(CVI) 10%

### 5. Esters of Dibasic Acids

The use of dibasic esters in this general condensation has also been studied in considerable detail. Much of this information has been disclosed in the form of patents. Two or more moles of biguanide react with esters of aliphatic dibasic acids (*700*) to yield diguanamines as expected. Insufficient biguanide gives rise to mixtures, said to consist of the polymethylene diguanamines and an ω-carbalkoxy-polymethyleneguanamine.

Esters of the lowest straight-chain dibasic acids (e.g. malonic acid) and biguanide are reported to yield almost exclusively the biguanide salt of the carboxymonoguanamine, the biguanide condensing with *one* of the ester groups only. This tendency to form the carboxymonoguanamines decreases with increasing molecular weight of the dibasic acid, and with an increased excess of biguanide over the theoretical requirement (*700*). As usual, the use of condensing agents (e.g. AlkONa) is generally desirable when the reaction is extended to alkyl- or arylbiguanides.

According to *Thurston* (*702*), the lower dibasic esters yield pure carboxymonoguanamines so readily that this reaction is an excellent synthetic route, when confined to the lower esters. It has been applied to 1-aryl- and alkyl-biguanides as usual, and to the alkali metal salts of half esters. The reaction involving diethyl malonate is represented below; succinic acid esters (*292, 699*) react similarly.

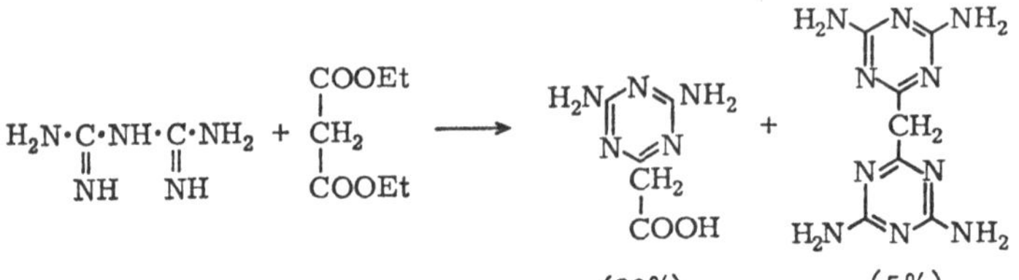

$H_2N{\cdot}C(=NH){\cdot}NH{\cdot}C(=NH){\cdot}NH_2$ + $CH_2(COOEt)_2$ ⟶ (diaminotriazinyl)$CH_2COOH$ (80%) + bis(diaminotriazinyl)$CH_2$ (5%)

The reaction has been reinvestigated thoroughly by *Sokolovska* et al. (*643*). Phenylbiguanide and diethyl malonate in presence of sodium ethoxide give in fact a mixture of four products (CVII-CX; x = 1) the structures of each of which were verified by synthesis. Phenylbiguanide and monoethyl malonyl chloride gave the same four compounds but in different ratios, with (CVII) and (CVIII) predominating. The action of higher dibasic esters studied by the same workers (*644*) gave comparable products (CVII, CX) (see also ref. *684*).

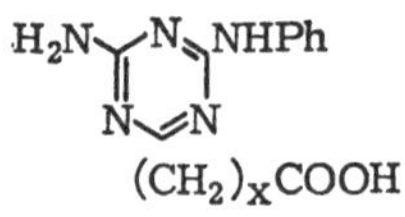

(CVII) 14·4% — $H_2N$, NHPh-triazinyl-$(CH_2)_xCOOH$

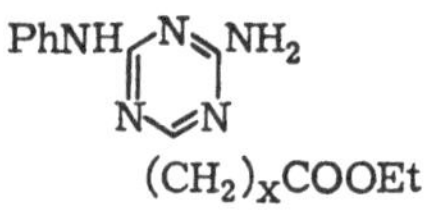

(CVIII) 22·7% — PhNH, $NH_2$-triazinyl-$(CH_2)_xCOOEt$

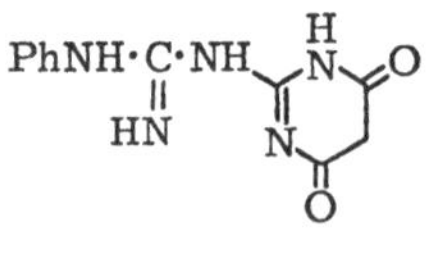

(CIX) 34·7% — $PhNH{\cdot}C(=HN){\cdot}NH$-pyrimidinedione

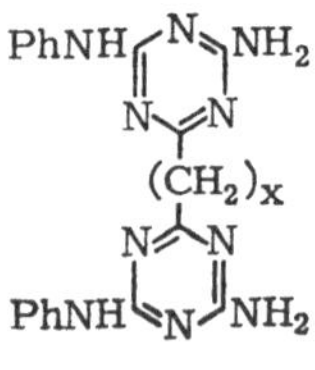

(CX) 9·2% — bis(PhNH, $NH_2$-triazinyl)$(CH_2)_x$

The replacement of one ester group by an acid chloride group increases the total yield of the ester (CVIII) and its parent acid (CVII) (to 70–89%). The yield of the monotriazine was also substantially improved by the continuous removal of the water from the reaction mixture. The reaction rate and yield of (CVIII) rose with increasing length of the alkyl chain of the half ester chloride. In view of these detailed observations, a re-evaluation of other patented examples of this ester-biguanide condensation might be desirable.

Aromatic dicarboxylic esters (or alkali salts of the half esters) in conjunction with two moles of biguanide (*710*) or its alkyl- or aryl-homologues afford the biguanide salt of an arylcarboxyguanamine, together with occasional traces of the appropriate diguanamine. Dibasic esters incorporating a double bond adjacent to the ester grouping (e.g. MeOOC·CH=CH·COOMe) react analogously (*293, 711*).

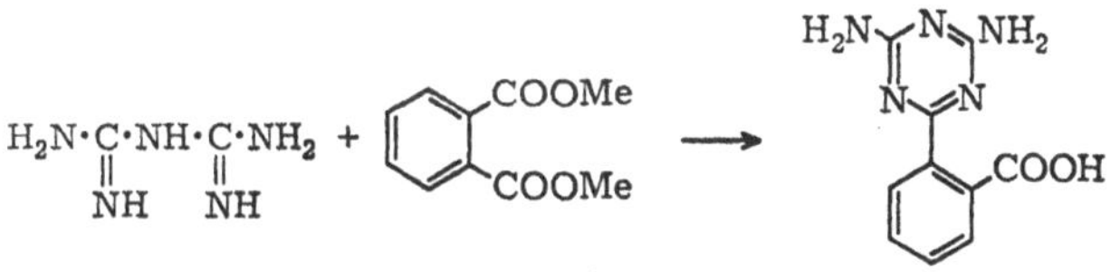

$$H_2N\cdot C(=NH)\cdot NH\cdot C(=NH)\cdot NH_2 + C_6H_4(COOMe)_2 \longrightarrow$$ (diaminotriazinyl)benzoic acid

The reaction of the "cyclic" biguanide o-phenylenebiguanide (CXI) with esters proceeds under the usual conditions (*537*); the biguanide appears to act almost entirely as an amidine, giving rise to the pyrimidine (CXII). Phenylbiguanide similarly affords (CXIII).

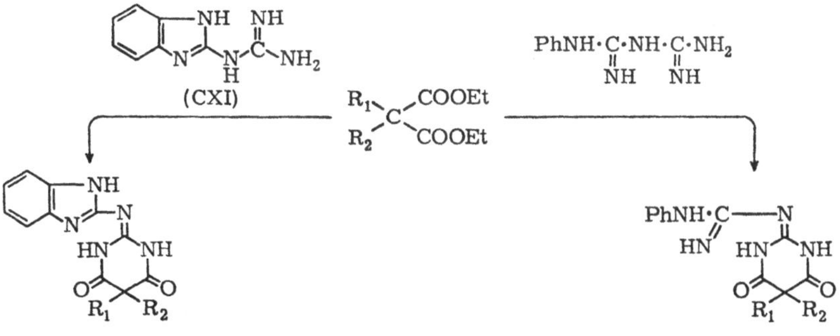

(CXI) + $R_1R_2C(COOEt)_2$ → (CXII)

$PhNH\cdot C(=NH)\cdot NH\cdot C(=NH)\cdot NH_2$ + $R_1R_2C(COOEt)_2$ → (CXIII)

The condensation of diethyl formylsuccinate with amidines is an established pyrimidine synthesis (*16, 714*). Extended to 1-substituted biguanides, the reaction yields 2-guanidino-4-hydroxy-5-carbethoxymethylpyrimidines as follows (*684*):

$$RNH \cdot \underset{\underset{NH}{\|}}{C} \cdot NH \cdot \underset{\underset{NH}{\|}}{C} \cdot NH_2 + HOCH{=}\underset{\underset{CH_2-COOEt}{|}}{C}-COOEt \longrightarrow RNH \cdot \underset{\underset{NH}{\|}}{C} \cdot NH\text{-}C_4HN_2(OH)CH_2COOEt$$

Carbamylguanamines (CXIV) may be prepared from biguanides and ester amides of dibasic acids in the usual manner (*709*), preferably in the presence of metal alkoxide. The rather difficultly accessible ester amides may be replaced by the more readily available imides of the dibasic acids (*454*).

$$H_2N \cdot \underset{\underset{NH}{\|}}{C} \cdot NH \cdot \underset{\underset{NH}{\|}}{C} \cdot NH_2 + \text{phthalimide} \longrightarrow \text{2,4-diamino-6-}(C_6H_4 \cdot CONH_2)\text{-s-triazine} \qquad (CXIV)$$

*Oxalates.* The interaction of phenylbiguanide and diethyl oxalate presents certain anomalous features (*485*). In methanol at 10°, the reactants produce an immediate quantitative precipitate of a yellow compound "Y", of composition $C_{10}H_9N_5O_2$. Continued reaction at room temperature slowly yields 2-metoxycarbonylphenylguanamine (CXV) and is complete after six days. These experimental results were confirmed independently by Italian investigators (*536*).

The structure of the labile intermediate compound "Y" is uncertain. On the basis of its physical properties and chemical behaviour (which cannot be detailed here), possible formulations that were originally (*485*) considered included a six-membered bridged carbonyl-structure, as well as seven- or five-membered rings. The Italian group of workers (*536*) favoured the seven-membered ring-structure (CXVI) for the intermediate "Y", and 5-phenylguanamine structures (CXXI) for the final stable products.

However, a reinvestigation by Russian workers (*642*) showed conclusively, that the final triazines were in fact substituted 6-phenylguanamines (CXV). Compound "Y" was represented as either CXVII or CXVIII, and its transformation into 2-amino-4-anilino-s-triazine-6-carboxylic acid (or a derivative) was explained in terms of the rupture of a —CO—NH— linkage (in either CXVII or CXVIII): the resulting 1-phenyl-4-acylbiguanide (CXIX) gave the final stable triazine-compound (CXX) on cyclisation.

In the presence of sodium ethoxide, the intermediate is not formed, and the reaction affords 2-amino-4-anilino-*s*-triazine-6-carboxylic acid in one stage (*642*).

The corresponding reaction of cyclohexylbiguanide has given comparable results (*607*). The primary intermediate was not identified but, as a basis for discussion, was formulated as the six-membered bridged carbonyl compound (CXXII).

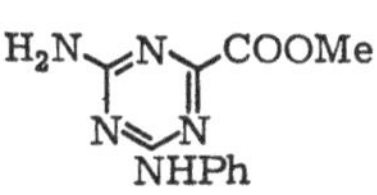

(CXV) 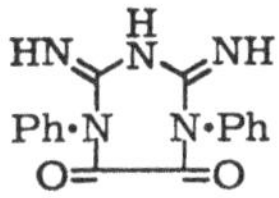 (CXVI) 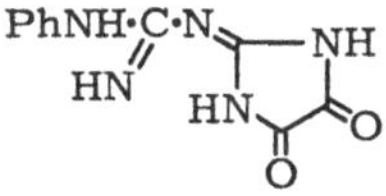 (CXVII) 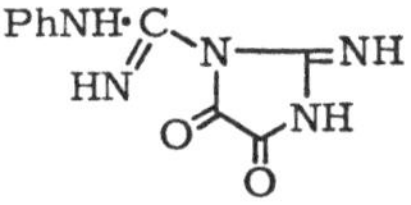 (CXVIII)

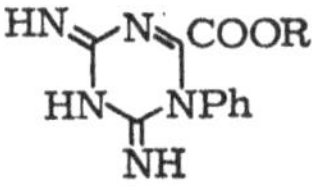

(CXXI)  (CXX) 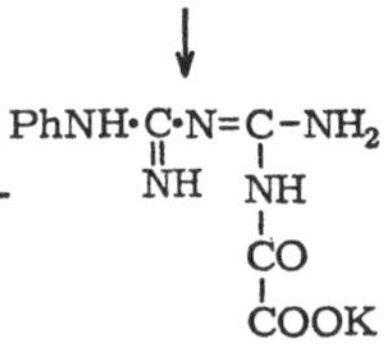 (CXIX) 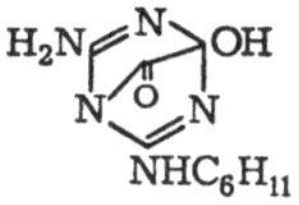 (CXXII)

## I. Action of Acids, Acid Amides, Halides and Anhydrides

Guanamines are obtainable from biguanides by the action of reagents other than esters. These include the amides, halides and anhydrides of organic acids, as well as the acids themselves.

### 1. Acid Amides

Formoguanamines are reported (*476*) to be readily accessible in high yields from biguanides and formamide (the latter acting both as solvent and reactant), the products being more easily purified than those obtained from the ester-biguanide condensation. Substituted biguanides react equally readily so that condensing agents are not required.

In the hands of other workers (*485*), however, the method has given less favourable results, affording much lower yields, even on prolonged refluxing in formamide. The product from phenylbiguanide and formamide was formulated as 3-phenyl- (*536*) and not 6-phenylguanamine (compare ref. *476*).

Boiling dimethylformamide gives (*485*) the same formoguanamines as does formamide; below 100° however, this compound does not react with biguanides and is in fact an excellent solvent for a variety of their condensations with other reactants (*375–378*).

Unlike acylamides, sulphonamides react with biguanides to yield mainly open-chain compounds, but the biguanide structure is cleaved in the process. Thus β-phenethylbiguanide and p-aminobenzenesulphonamide yield a mixture of p-aminobenzenesulphonylguanidine [$RSO_2NH \cdot C(:NH)NH_2$] and 1-p-aminosulphonyl-3-β-phenethylguanidine [$RSO_2NH \cdot C(:NH)NHCH_2CH_2Ph$] (15 mins. at 50–55° in methanol); a similar reaction ensues with 1,1-dimethylbiguanide.

## 2. Acids, Acid Halides and Anhydrides

Boiling 96% formic acid converts biguanides into formo-guanamines (*158*, *485*): this variation of the synthesis has been widely used, especially in patent work (*19*, *113*, *127*, *221*, *593*, *599*, *677*).

Treatment of biguanides with acyl halides in alkaline media also yields guanamines (*518*). The use of sodium hydroxide results in only low (10–15%) yields of difficultly purifiable guanamines (*518*), but the reaction proceeds smoothly and in good yield (*518*) in milder alkalis (e.g. sodium carbonate). In the presence of metal alkoxides (*291*) the reaction may be carried out at room temperature. These striking improvements are also effective with substituted biguanides; the method is generally applicable to the synthesis of a wide range of guanamines (CXXIII), including those incorporating higher alkyl-groups in the 2-position (*700*). Substituted carboalkoxy-acid halides (*292*) and unsaturated acid halides (*293*) afford the corresponding carboxyguanamines or unsaturated guanamines respectively.

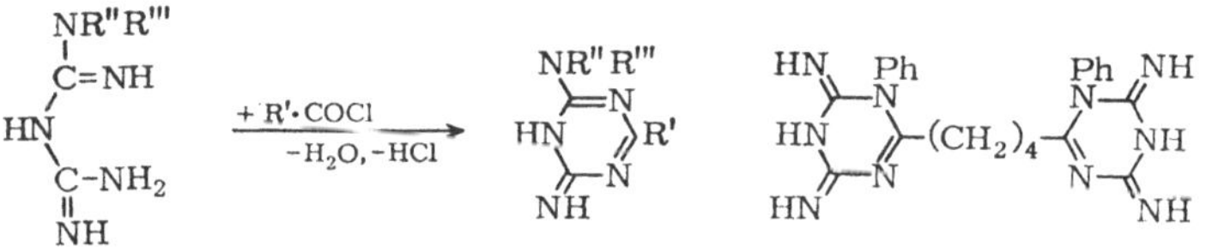

$$HN(C(=NH)NR''R''')(C(=NH)NH_2) \xrightarrow[-H_2O,\,-HCl]{+R'\cdot COCl} \text{(CXXIII)}$$

(CXXIII) (CXXIV)

Phenylbiguanide and stearoyl chloride are reported (*236*) to yield 5-phenylformoguanamine, and not 5-phenyl-6-heptadecylformoguanamine, as would be expected from the general course of this reaction. The product of phenylbiguanide and adipyl chloride has been formulated (*237*) as the 5-phenyl- rather than the 6-phenyl-ditriazine.

An important *industrial improvement* was provided by the observation (*455*) that under appropriate conditions cheap salts of biguanide could be substituted for the costly free bases. Thus acid chlorides or anhydrides are claimed to react in inert organic solvents (e.g. acetone, dioxan, pyridine, benzene) with biguanide salts in the presence of alkalis, giving near-quantitative yields of guanamines. Substitution of an acid chloride or anhydride by an ester in non-hydroxylic solvents gives only negligible yields. The reaction, which may be extended to 1-substituted biguanides, offers considerable economic advantages (*455*). As an example, biguanide sulphate and succinic anhydride in dioxan-sodium hydroxide provide 62% yields of 2,4-diamino-6-ω-hydroxyethyl-*s*-triazine. Absence of dioxan or use of ethanol as solvent affect the yield adversely (5%).

3. Action of other Reagents

Phenylbiguanide, when boiled with either diformylhydrazine or triethoxymethane gave a product formulated (*536*) as 3-phenylguanamine. Since the authors synthesized this "3-phenylguanamine" by alternative methods, several of which have since been shown (*642*) to yield 6-phenylguanamine, it is likely that the above product is in fact 6-phenylguanamine.

## J. Reactions with Keto-Compounds

The interaction of biguanides and ketones was studied thoroughly by *Curd*, *Rose* and their co-workers (*58*), and independently by *Modest* and *Levene* (*443*). It has become a widely used synthetic route to certain dihydrotriazines (*27, 43, 44, 56, 136, 156, 185, 200, 226, 417, 471, 493, 534*), the chemistry of which has been reviewed recently (*442*).

1. 1-Substituted Biguanides

p-Chlorophenylbiguanide was found (*58*) to react with acetone in the presence of piperidine as a catalyst to yield a product which did not give a copper complex and was formulated as 2-amino-4-p-chloroanilino-6-dimethyl-5,6-dihydro-s-triazine (CXXV). Since 1-aryl-2,5 (or 4,5)-dialkylbiguanides, as well as 2-p-chlorophenylguanidino-4-β-diethylamino ethylamino-6-methylpyrimidine (*147*) retain the power of forming copper complexes, the alternative formulation of the above product as

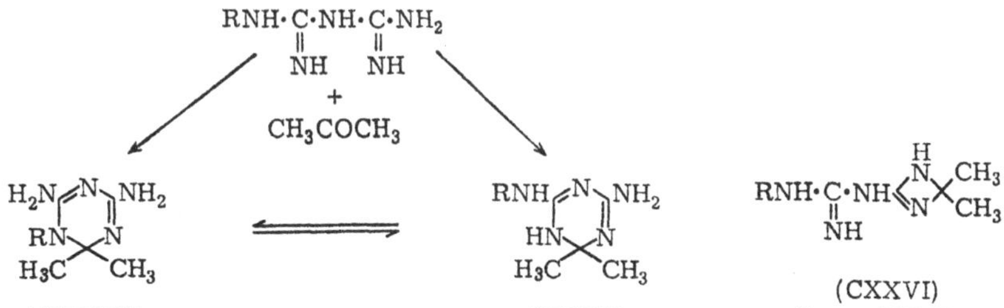

RNH·C(=NH)·NH·C(=NH)·NH$_2$ + $CH_3COCH_3$

(CXXVII) ⇌ (CXXV)

(CXXVI)
R = p-$ClC_6H_4$·

(CXXVI) may be rejected. Structure CXXVI is also inadmissible because the product cannot be reduced to 1-p-chlorophenyl-5-isopropylbiguanide, and has no antimalarial properties.

Further details concerning this reaction were provided by *Carrington, Crowther* and *Stacey* (*95*) in investigations concerning the structure of the biologically active metabolite of Paludrine. The condensation of p-chlorophenylbiguanide hydrochloride and acetone in aqueous solution gave small amounts of 4,6-diamino-1-p-chlorophenyl-1,2-dihydro-2,2-dimethyl-s-triazine (CXXVII). The yield depends, *inter alia*, on the amount of acetone used and becomes near-quantitative in the presence of a small excess of acid. *Modest* et al. (*443*) have later suggested that under neutral conditions the reaction does not proceed at all, and that the minute yields of the dihydrotriazine (CXXVII) had been due to the catalytic effect of a trace of acid as impurity in the reaction mixture. The condensation is reversible, the ketone and biguanide being regenerated by hot dilute aqueous acid (*95*).

The experimental procedure can be varied by the use of reagents which will give biguanides *in situ* and thus form the required triazine. Examples include (*493, 554*) the interaction of arylamines (e.g. p-chloroaniline hydrochloride) and cyanoguanidine, followed by acetone and hydrochloric acid. Acetone may be replaced by its bisulphite compound, by diethyl acetal or by isopropenyl acetate, but p-chlorophenylbiguanide fails to condense (*113*) with acetone diethyl acetal at 130°. Under the correct acid conditions, the isomeric p-chloroanilino-triazine (CXXV) is not formed; on the other hand, the free p-chlorophenylbiguanide base did not react with acetone in the absence of a catalyst.

The 1-chlorophenyl-compound (CXXVII) isomerises readily to its p-chloroanilino-isomer (CXXV) on being heated above its melting point, or warmed in solution. The two triazines are easily distinguished by their ultraviolet absorption spectra. In view of the ease of this isomerisation, analogues of (CXXVII) obtained by this route must be examined closely to ascertain their true nature.

(CXXVIII) $H_2N$, N, $NH_2$, RN, N, Me, Et $\xrightarrow{OH^-}$ RNH, N, $NH_2$, HN, N, Me, Et (CXXIX)

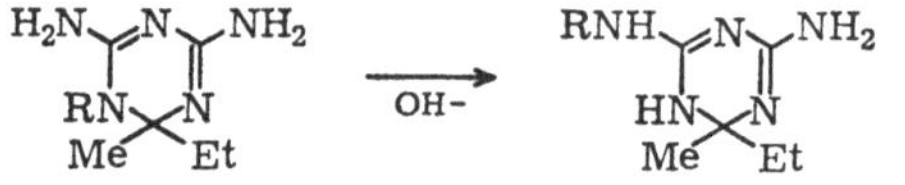

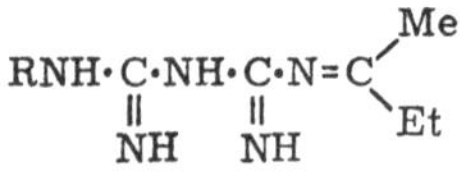

RNH·C(=NH)·NH·C(=NH)·N=C(Me)(Et)

(CXXX)

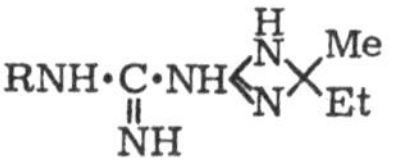

RNH·C(=NH)·NH–(N(H), N, C(Me)(Et))

(CXXXI)

RNH·C(=HN)–N(N, $NH_2$, C(Me)(Et))

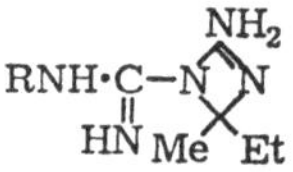

(CXXXII)

Extension of this condensation to mixed ketones (*95*) provided further confirmation for the above interpretation. Thus, methyl ethyl ketone and p-chlorophenylbiguanide gave successively the triazine derivatives (CXXVIII) and (CXXIX). The latter (CXXIX) was resolvable into its enantiomers by (+)-tartaric acid; this observation supports the six-membered ring structure of (CXXIX) and, because of the very gentle conditions of its isomerisation, that of its isomer (CXXVIII).

Of the other possible structures (*95*) of the condensation product, the Schiff's base (CXXX) is not resolvable, nor, because of the tautomeric nature of the guanidine system, is the four-membered ring compound (CXXXI). A compound of the improbable structure (CXXXII) should yield a copper complex, which the condensation products failed to do.

The condensation of 2-naphthylamine hydrochloride, cyanoguanidine and acetone yields, in addition to the expected dihydro-s-triazine, 1-(2-naphthyl)biguanide and 3-guanidino-1-methylbenzo[f]quinazoline (*554*).

## 2. 1,1-Disubstituted Biguanides

In the general reaction, 1,1-disubstituted biguanides (*95*) (e.g. 1-p-chlorophenyl-1-methylbiguanide) afford the same product (CXXXIII) under both acidic and basic conditions. The phenyl analogue gave similar results (*95*). Independent studies by *Modest* et al. (*443*) confirmed the assigned structure (CXXXIII).

## 3. 1,2-Disubstituted Biguanides

The interaction of 1,2-disubstituted biguanides and acetone has provided particularly significant results (*95*). 1-p-Chlorophenyl-2-methylbiguanide and acetone reacted in the expected manner, under acidic *or* basic conditions respectively, to yield the isomers (CXXXIV) *or* (CXXXV).

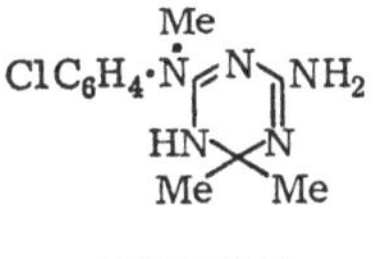

(CXXXIII) 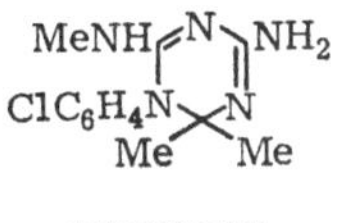 (CXXXIV) 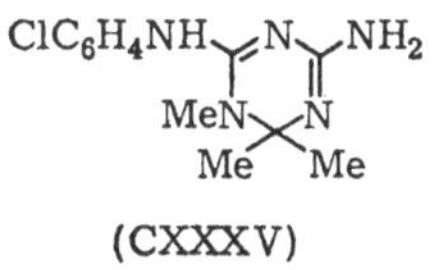 (CXXXV) 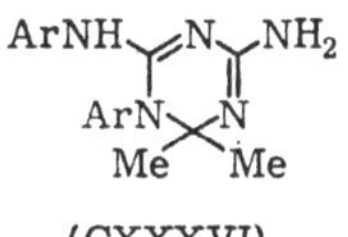 (CXXXVI)

Isomerisation of the acid-catalysed condensation product (CXXXIV) gave, according to the ultraviolet spectral evidence, equilibrium mixtures of the two isomeric forms. This view was strengthened by the observation that treatment with alkali of the "base-catalysed" condensation product (CXXXV) gave a similar equilibrium mixture [3-4 parts of (CXXXV) to one of (CXXXIV)].

The observations suggest that the two condensations are direct and distinct. Had the base-catalysed reaction first given the aryl-dihydrotriazine (CXXXIV) which, under the prevailing basic conditions rearranged to the anilino-dihydro-triazine (CXXXV), an equilibrium mixture of the two would have been expected. Since, however, the pure anilino-dihydrotriazine (CXXXV) was isolated, its formation is considered to be direct and independent of that of its isomer (*95*). The formation of (CXXXIII) from 1-aryl-1-alkylbiguanides (see preceding Section), which cannot proceed by way of an aryl-dihydrotriazine, bears out this conclusion.

The condensation of acetone and 1,2-diarylbiguanides bearing identical substituents yields the same alkali-stable product (CXXXVI) under both acidic and basic conditions (*95*).

### 4. Tri- and Higher Substituted Biguanides

6-Amino-1-p-chlorophenyl-4-dimethylamino-1,2-dihydro-2,2-dimethyl-s-triazine (CXXXVII) was obtained from acetone and 1-p-chlorophenyl-5,5-dimethylbiguanide under acid conditions, and was isomerised by alkali to 6-p-chloroanilino-4-dimethylamino-1,2-dihydro-2,2-dimethyl-s-triazine (CXXXVIII), which also arose directly under basic conditions (*95*). Piperidine was found useful in promoting an analogous condensation (*534*).

$Me_2N$, $NH_2$, $N{\cdot}C_6H_4Cl$, Me, Me

(CXXXVII)

$Me_2N$, $NHC_6H_4Cl$, NH, Me, Me

(CXXXVIII)

$ClC_6H_4{\cdot}NH{-}C(=N{\cdot}C(=NH){\cdot}NH_2){-}NMe_2$

(CXXXIX)

$ClC_6H_4NH$, NHMe, NMe, Me, Me

(CXL)

Acetone failed to react with 1,2,2-trisubstituted biguanides (specifically with compound CXXXIX), both in acidic and basic media. Although no reaction would be expected in this case in a basic environment, the failure under acidic conditions is unexpected.

No reaction occurred with 1,4,5-trisubstituted biguanides (e.g. 1-p-chlorophenyl-4,5-dimethylbiguanide) and acetone in basic media. Prolonged action under acid conditions gave small amounts of a picrate which appeared to correspond to the triazine (CXL).

1-p-Chlorophenyl-2,4,5-trimethylbiguanide failed to react both in acidic and basic media. In view of the results of the preceding example, however, this is not surprising.

*Summary*

Several of the above results were arrived at independently by *Modest* and *Levene* (*443*) in their work on the active metabolite of Paludrine. They confirmed the formation of aryldihydrotriazines from 1-aryl-1-methyl- and 1-aryl-4,5-dimethylbiguanides; they also extended the synthesis to 1-aryl-5-methyl- and 1-aryl-5,5-dimethylbiguanides. Additional confirmatory results were provided by the independent researches of *Loo* (*397*) who has proposed a mechanism for the base-catalysed isomerisation of the aryldihydrotriazines of type CXXVII.

The reaction is best carried out in the appropriate anhydrous aldehyde or ketone, which thus functions as reactant and solvent, or in ethanol, in the presence of an optimum of 1.5 equivalents of strong acid per equivalent of arylbiguanide (base). Acetic acid as a catalyst produces, anomalously, the "base-catalysed" anilino-triazines (*126, 443*). The reaction time varies from one hour to two weeks. The rates diminish in the following order:

Aliphatic Aldehydes > Aromatic Aldehydes > Ketones

Aldehydes are generally suitable, excepting highly reactive examples such as formaldehyde which may undergo co-polymerisation with the arylbiguanide in preference to dihydrotriazine-formation.

a) *Ketones*

The ease with which ketones participate in this synthesis is comparable with their power of forming bisulphite-addition products. Thus, the reaction goes to completion with acetone, cyclopentanone and cyclohexanone, but is progressively inhibited with higher aliphatic ketones in which the larger adjacent groups exert an increasingly effective steric hindrance on the carbonyl-group. Methyl ethyl ketone behaves anomalously, leaving some unreacted biguanide. Aryl ketones (*131*, *443*) fail to react.

Table 9. *Interaction of substituted biguanides with acetone*

| | | |
|---|---|---|
| 1-Aryl-5-methylbiguanide + acetone | ⟶ | dihydrotriazine |
| 1-Aryl-5,5-dimethylbiguanide + acetone | ⟶ | dihydrotriazine |
| 1-Aryl-1-methylbiguanide + acetone | ⟶ (crossed) | dihydrotriazine |
| 1-Aryl-4,5-dimethylbiguanide + acetone | ⟶ (crossed) | dihydrotriazine |

The results summarised in Table 9 suggest that the presence of one hydrogen atom at each of the terminal nitrogen atoms of the arylbiguanide is essential to the success of this synthesis. The reaction is thus a condensation of the Mannich-type, involving the hydrogen atoms at $N^1$ and $N^4$ of the arylbiguanide, and the oxygen grouping of the carbonyl reagent.

A particularly interesting example is the fusion of cyclohexanone (*20*) with biguanide at 140°, which is reported to yield the spirane 2-cyclohexylguanamine (CXLII) by cyclisation of the intermediate (CXLI) involving a proton shift:

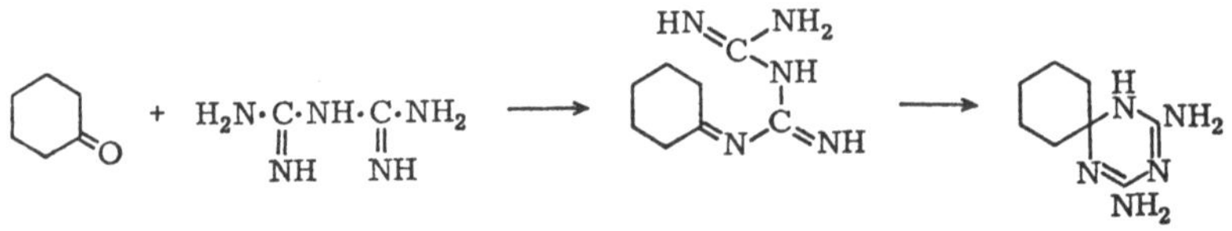

$$+ \; H_2N\cdot C(=NH)\cdot NH\cdot C(=NH)\cdot NH_2 \longrightarrow \text{(CXLI)} \longrightarrow \text{(CXLII)}$$

(CXLI) (CXLII)

The experimental procedure using biguanides generated *in situ* from arylamines and cyanoguanidine, which was briefly mentioned by *Carrington* et al. (*95*) has been studied in greater detail by *Modest* (*441*) (see also ref. *417*). This variation of the general synthesis has been fairly widely used in the production of dihydrotriazines (*34, 156, 555, 574*).

Studies have recently become available concerning the kinetics and mechanism of the isomerisation, in dilute aqueous solution, of dihydrotriazines of this type (e.g. 4,6-diamino-1-(3,5-dichlorophenyl)-1,2-dihydro-2,2-dimethyl-1,3,5-triazine), and of their degradation to substituted biguanides (*683*).

b) *Aldehydes*

Relatively few examples of the use of aldehydes in this general reaction have been investigated. Phenylbiguanide reacts with benzaldehyde under acid conditions to yield the dihydrotriazine (CXLIII) (*95*) (see also ref. *493*).

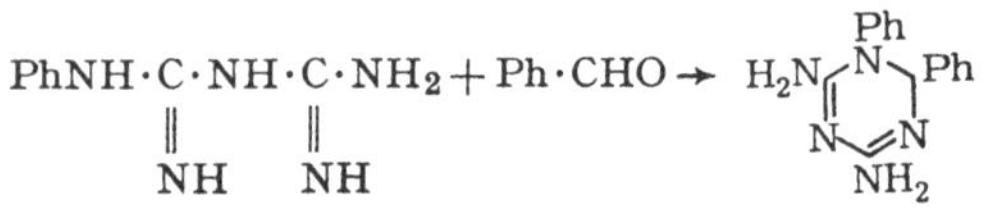

$$PhNH\cdot C(=NH)\cdot NH\cdot C(=NH)\cdot NH_2 + Ph\cdot CHO \rightarrow \text{(CXLIII)}$$

(CXLIII)

Other aromatic aldehydes and ω-phenylalkyl aldehydes react analogously (*27*). Chloral gives a complex reaction mixture (*485*), from which 6-phenylguanamine (11%) was isolated. p-Biguanidophenylarsonic acid is said (*408*) to add one mole of formaldehyde, but the structure of the addition product was not specified.

5. Miscellaneous Keto-Compounds

(1-p-Arsonophenyl)biguanide fails to react with ethyl acetoacetate, but condenses with acetylacetone with loss of two moles of water, to give the appropriate pyrimidine (CXLIV) exclusively. The reaction is of general applicability (*417*, *590*), affording for example, 2-(4-carboxyphenylguanidino)-4,6-dimethylpyrimidine from 4-biguanidobenzoic acid.

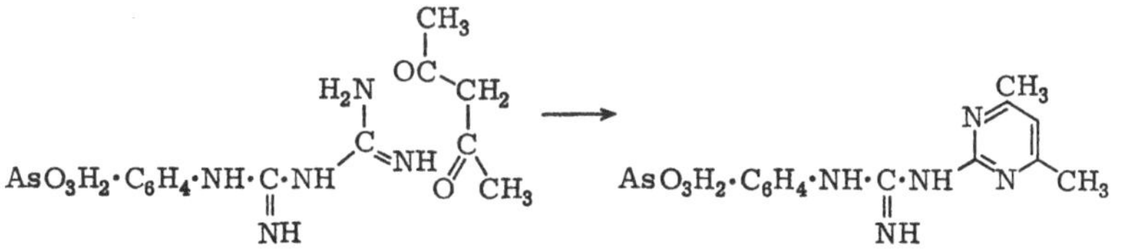

$AsO_3H_2 \cdot C_6H_4 \cdot NH \cdot C(=NH) \cdot NH \cdot C(NH_2)=NH$ + $CH_3COCH_2COCH_3$ → $AsO_3H_2 \cdot C_6H_4 \cdot NH \cdot C(=NH) \cdot NH$-(4,6-dimethylpyrimidin-2-yl)

(CXLIV)

$PhNH \cdot C(=NH) \cdot NH$-(5-phenylazopyrimidin-2-yl)

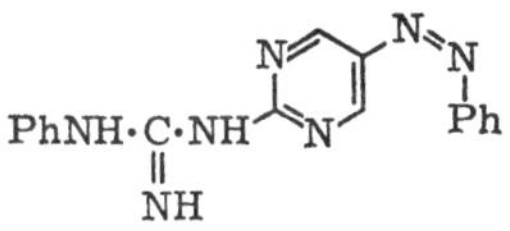

(CXLV)

Phenylazo-diketones (*547*) and dialdehydes (*423*) [e.g. PhN=N CH(CHO)$_2$] similarly yield 3-phenylazopyrimidine derivatives (e.g. CXLV).

c) *Chelidonic Acids*

The pronounced reactivity of biguanides is illustrated by their interaction with 3-phenylchelidonic acid (*466*) (CXLVI) which proceeds readily in good yields in boiling ethanol: the keto group, the carboxyl groups and the pyrone oxygen (of CXLVI) all participate in this reaction, giving rise to CXLVII (R = aryl(or alkyl)biguanidyl).

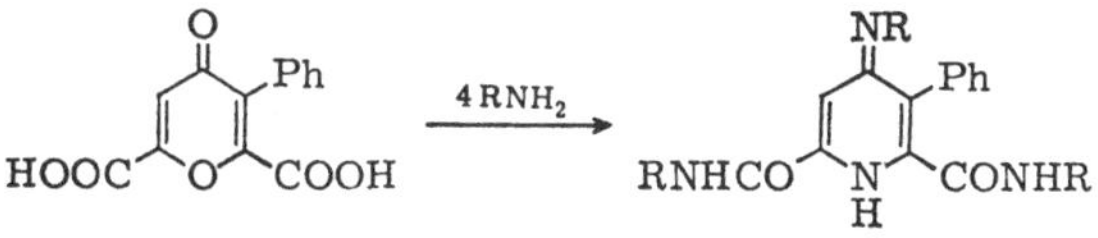

HOOC–(3-phenyl-4-pyrone)–COOH $\xrightarrow{4\,RNH_2}$ RNHCO–(4-NR-3-phenyl-1H-pyridine)–CONHR

(CXLVI) (CXLVII)

The same product is obtained from 3-phenylchelidamic acid, the pyrone oxygen of which is already substituted by imino-nitrogen, only three molecules of biguanide being required for the formation of the final product (CXLVII). Similar reactions occur (*464*) with the parent chelidonic acid and with γ-pyrone.

## K. Reactions with Cyanamides and Carbodiimides

The condensation of biguanide and cyanamide (which, in its diimino-form, may be regarded as the parent compound of carbodiimides (*373*)) has apparently not been studied. The patent literature (*85, 109*) describes the manufacture of melamine by heating mixtures of biguanide or guanidine or their salts with cyanamide or dicyandiamide to 140—240°, but since each of these compounds *alone* yields melamine under these conditions, the course of these reactions is not clear. The only valid example appears to be the condensation of p-chlorophenylcyanamide and p-chlorophenylbiguanide to 2,4-di(p-chlorophenyl)melamine (*304*).

The reaction between biguanides and carbodiimides in dimethyl-formamide provides a general route to substituted melamines (*375*). Thus, biguanide or its 1-mono- or 1,2-disubstituted homologues afford, respectively, satisfactory yields of mono- or 2,4-disubstituted melamines (CXLVIII) or 1,2,6-trisubstituted isomelamines (CXLIX). In each case, the reaction probably involves primary addition of the reactants, followed by the cyclisation of the resulting intermediate labile triguanide, with loss of amine (*375*):

$$R'NH\cdot C(=NH)\cdot NH\cdot C(=NH)\cdot NH_2 + RN{=}C{=}NR \longrightarrow R'NH\cdot C(=NH)\cdot NH\cdot C(=NH)\cdot NH\cdot C(=NR)\cdot NHR \xrightarrow{-RNH_2}$$

R'NH, NHR, $NH_2$-substituted s-triazine ring

(CXLVIII)

$$RNH\cdot C(=NR)\cdot NH\cdot C(=NH)\cdot NH_2 + RN{=}C{=}NR \longrightarrow RNH\cdot C(=NR)\cdot NH\cdot C(=NH)\cdot NH\cdot C(=NR)\cdot NHR \xrightarrow{-RNH_2}$$

RN, N-R, NR, HN, NH, NH isomelamine ring

(CXLIX)

## L. Reactions with Isocyanate and Isothiocyanate Esters

The interaction of biguanide (CL) and isothiocyanate esters in dimethyl-formamide under mild conditions (*376*) gives excellent yields of 1-substituted hexahydro-4,6-di-imino-s-triazine-2-thiones (CLII), together with small quantities of thioammeline (CLIII) and monosubstituted melamines (CLIV). Under more severe conditions, the latter (CLIV) become the main products. The production of these triazines is explained by a mechanism involving the primary formation of addition products (CLI), followed by their cyclisation with loss of either ammonia, hydrogen sulphide, or primary amine (*376*).

By an analogous mechanism, 1-phenylbiguanide (CLV) yields the corresponding more highly substituted melamines (mostly CLVII, some

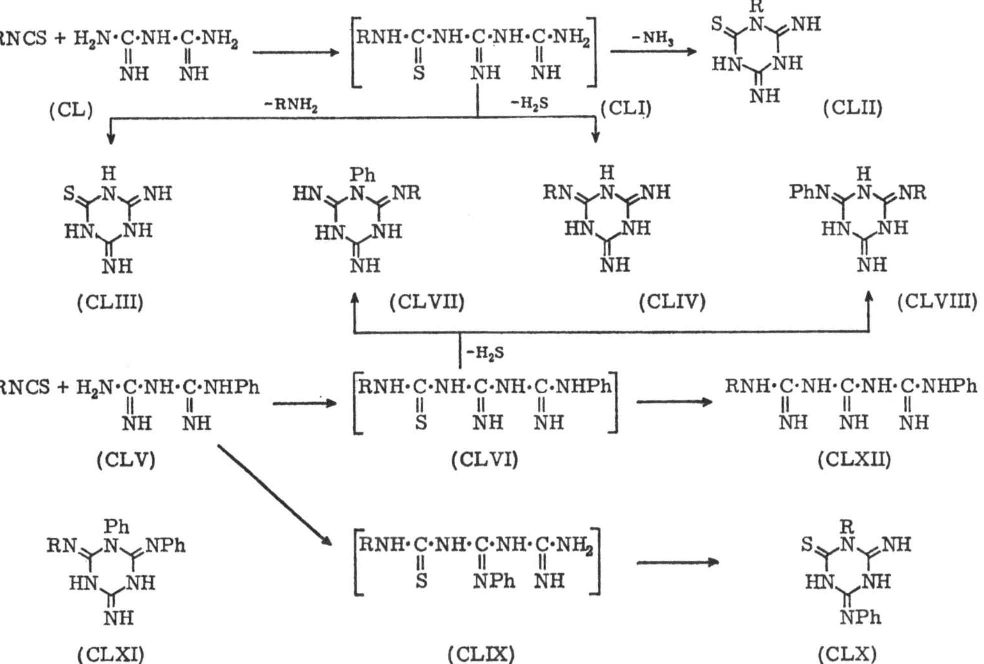

CLVIII), *or* thioammelines, depending on the reaction conditions; the available evidence suggests that the latter are 1-substituted hexahydro-6-imino-4-phenylimino-s-triazine-2-thiones (CLX) (*377*, *378*). Finally, 1,2-diphenylbiguanide produces, by the general reaction, 6-substituted 1,2-diphenylisomelamines (CLXI) exclusively, hydrogen sulphide being eliminated preferentially from the more highly substituted primary adducts (*377*).

The postulated non-cyclic intermediates (CLI, CLVI) appear generally to be too labile to be isolated and characterised (*376*, *377*) (see also ref. *647*). Claims (*297*, *557*, *581*) of the successful preparation of compounds of this structure (e.g. CLVI, R = Ph) (*297*) have so far not been substantiated (*377*). Their alleged conversion (*297*, *557*) into triguanides (CLXII) by ammonolysis must therefore also be regarded as unconfirmed (see also Section III K).

The condensation of isocyanate esters with diguanides proceeds in an entirely comparable manner, providing a corresponding series of s-triazin-2-ones (i.e. substituted ammelines) (*378*). However, since loss of water from carbamoyl-intermediates [e.g. RNHCONH·C(:NH)NH·C(:NH)-$NH_2$] occurs much less readily than loss of hydrogen sulphide from their thiocarbamoyl-analogues, melamines are not formed in this reaction (*378*). The production of adducts from phenyl isocyanate and tetra-, penta-, or hexamethylbiguanide in dry benzene has been described (*211, 212*). These phenylcarbamoyl-adducts have a catalytic effect on urethane-formation from phenyl isocyanate and alcohols (*211*).

## M. Reaction with Diazonium Salts

Isopropylbiguanide reacts additively with p-chlorophenyldiazonium chloride (*517*) in the expected manner:

$$\text{p-ClC}_6\text{H}_4\cdot\text{N}{=}\text{N}\cdot\text{Cl} + \text{H}_2\text{N}\cdot\underset{\underset{\text{NH}}{\|}}{\text{C}}\cdot\text{NH}\cdot\underset{\underset{\text{NH}}{\|}}{\text{C}}\cdot\text{NHPr}^{\text{i}}$$

$$\longrightarrow \text{p-ClC}_6\text{H}_4\text{N}{=}\text{N}\cdot\text{NH}\cdot\underset{\underset{\text{NH}}{\|}}{\text{C}}\cdot\text{NH}\cdot\underset{\underset{\text{NH}}{\|}}{\text{C}}\cdot\text{NHPr}^{\text{i}}$$

The resulting 1-(p-chlorophenylazo)-5-isopropylbiguanide is cleaved by hydrochloric acid to p-chlorophenol and 1-isopropylbiguanide. Similar observations concerning the reaction of diazonium salts with biguanides have appeared in the patent literature (*403*).

## N. Halogenation

The halogenation of 1-arylbiguanides has been studied by *Curd, Rose* and co-workers (*134, 135*). On treatment of the base or its hydrochloride in acetic or concentrated sulphuric acid with the calculated quantity of chlorine, bromine or iodine chloride, the biguanide-moeity is not affected: halogenation occurs in the aromatic ring, and obeys the usual rules of aromatic substitution (*134, 135*). The introduction of alkyl groups (at $N^2$ and $N^4$ in the biguanide structure) has little effect on the course of halogenation (*134*).

A later patent (*307*) has claimed the successful chlorination of biguanides in positions $N^2$ and $N^4$, preferably using 1,5-di- or tetra-substituted biguanides. The reaction is performed by adding the biguanide to aqueous hypochlorous acid (preferably in 10—15% excess). Two chlorine atoms are successively introduced, the first at 0—10°, and the second at higher temperatures (not exceeding 30°), although some dichlorination tends to occur at lower temperatures.

## O. Reactions with Boron Compounds

Biguanide reacts with aminoboranes and boron acids to give heterocyclic compounds (diaza-azonia-boronides) containing tetravalent boron (*436*).

Bis(dimethylamino)phenylborane in hot pyridine, for example, yields dimethylamine and a polymeric form (CLXIV) of the expected heterocycle (CLXIII).

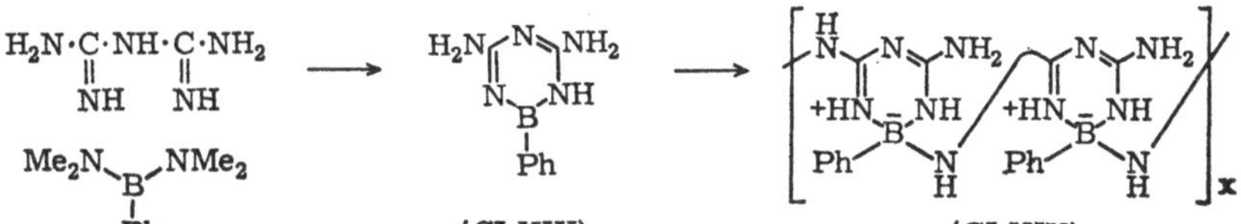

(CLXIII) (CLXIV)

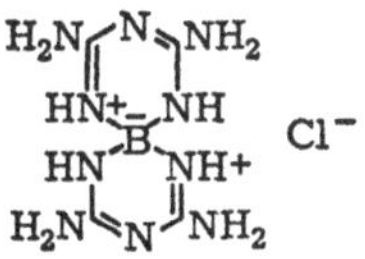

(CLXV) 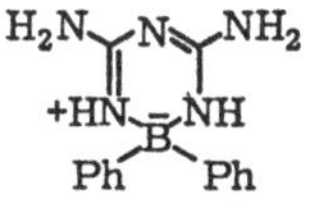 (CLXVI) 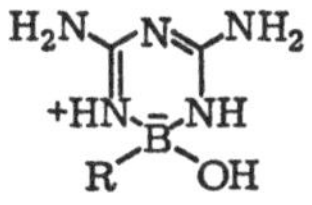 (CLXVII)

Tris(dimethylamino)borane similarly fails to yield a trivalent boron heterocycle, but gives the spiro-compound (CLXV). A monocyclic system of type (CLXVI) does arise from diethylaminodiphenylborane, and of type (CLXVII) from phenylboronic acid and its analogues (*436*).

## P. Miscellaneous Addition Reactions

### 1. Reaction with Ethylene Sulphide and Ethylene Oxide

According to the patent literature (*446, 447*), biguanide reacts additively with boiling ethylene sulphide (employed in two molar excess) to yield mercaptoethylbiguanide (CLXVIII; X = S).

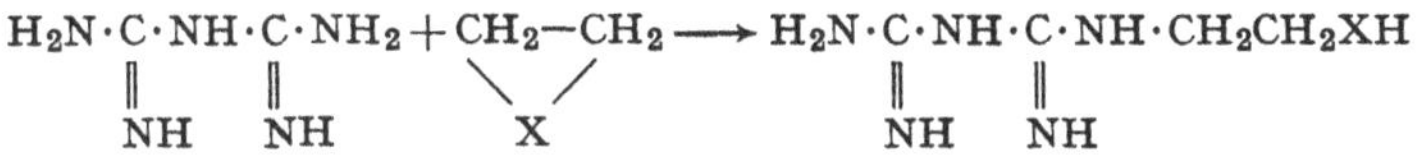

$$H_2N\cdot C(=NH)\cdot NH\cdot C(=NH)\cdot NH_2 + CH_2{-}CH_2(X) \longrightarrow H_2N\cdot C(=NH)\cdot NH\cdot C(=NH)\cdot NH\cdot CH_2CH_2XH$$

(CLXVIII)

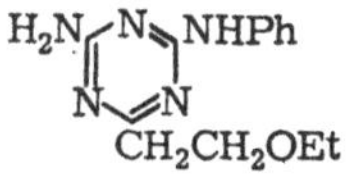

(CLXIX)

Similar reports (*290*) refer to adducts (presumably of the same type, CLXVIII; X = O) between biguanides of high molecular weight and alkylene oxides.

### 2. Complex Formation with Guanamines

The interaction of phenylbiguanide and ethyl β-ethoxypropionate in the presence of sodium methoxide yields a readily dissociable molecular complex of the expected guanamine and phenylbiguanide (*486*) (CLXIX). Aqueous picric acid cleaves the complex into the picrates of its constituents. However, the complex is not formed from the pre-formed 2-β-ethoxyethyl-6-phenylguanamine (or its methyl analogue) and phenylbiguanide in methanol or acetonitrile, and its nature is not completely understood (see also ref. *605*).

### 3. Adducts with α,β-Unsaturated Compounds

Hexamethylbiguanide (CLXIX) reacts additively in acetonitrile with α,β-unsaturated compounds activated by an electron-withdrawing group (e.g. with vinyl cyanide) to yield saturated adducts (CLXX). By a suitable choice of reagents, adducts bearing substituents other than the nitrile group (e.g. carbamoyl, carbalkoxy) are accessible (*213*):

$$\underset{\text{(CLXIX)}}{Me_2N\cdot \underset{\underset{NH}{\|}}{C}\cdot N{=}\underset{\underset{NMe_2}{|}}{C}\cdot NMe_2} \xrightarrow{CH_2:CH\cdot CN} \underset{\text{(CLXX)}}{Me_2N\cdot \underset{\underset{NCH_2CH_2CN}{\|}}{C}\cdot N{=}C(NMe_2)_2}$$

## Q. Promoting Influence of Biguanide on Other Reactions

The reactions of acetylene, including its polymerisation, that occur in the presence of cupric chloride solution have been studied (*670*) in the (additional) presence of biguanide dihydrochloride. This solution has the unique property of promoting the formation of acetic acid from acetylene.

# VIII. Estimation of Biguanides

Since biguanides are of outstanding medicinal importance (see Section IX), several methods have been devised for their estimation both in the pure state, and in body fluids (e.g. blood, urine, etc.). The use of picrolonic acid in identifying drugs of basic character, including biguanides, has been described (*759*).

Paludrine in bulk may be estimated by titration with perchloric acid in glacial acetic acid (*653*), or gravimetrically as its copper complex

(*653*), or Reineckate (*645*). The copper complex method is applicable (*228*) to estimating small amounts (5—100 mg. per litre) of biguanides; if the copper complex is soluble in a hydrocarbon solvent, a colorimetric technique may replace the gravimetric method (*355*, *648*). The nickel complex may be similarly used (*355*, *648*) both gravimetrically and colorimetrically.

One of the earliest methods of estimating Paludrine and related compounds in blood was their pressure hydrolysis (*652*, *653*) to the amine followed by diazotisation, coupling, and colorimetric estimation of the resulting dye. Other colorimetric estimations of Paludrine utilise its reaction with eosin (*48*) or bromothymol blue (*360*), or a modified Sakaguchi reaction involving potassium hypobromite and isopropanol (α-naphthol being unnecessary) (*487*). Several qualitative tests which rely on the shape of crystals formed by Paludrine and related compounds with certain reagents have been given (*111*).

β-Phenethylbiguanide may be estimated by a colorimetric method involving α-naphthol, diacetyl propanol and sodium hydroxide (*218*, *610*) which allows the detection and estimation of $5 \times 10^{-6}$g. of the compound (*218*).

1,1-Dimethylbiguanide may be estimated by colorimetric procedures based on its reaction with diacetyl in presence of α-naphthol (*618*), or a modified Sakaguchi (*343*, *508*) and other (*46*) reactions.

A procedure for estimating hexamethylene 1,6-bis[1-(5-p-chlorophenyl)biguanide] ("chlorohexidine") in the presence of anesthesine in tablets has been described (*617*). $N^1$,$N^1$Anhydro[bis(β-hydroxyethyl)]-biguanide (*616*) and other biguanides (*137*) have been determined by spectrophotometric (*137*, *616*) (including infra-red (*563*)) techniques. Chromatographic procedures have also been described (*26*, *449*, *467*).

Biguanide sulphate has been proposed (*525*) as a primary standard in acidimetry. It has the advantage that it is cheaper and more stable than some other primary reagents.

## IX. Medical Uses of Biguanides and Dibiguanides

### A. Biguanides

Biguanides have been used for a variety of purposes in medical practice but their clinical importance is undoubtedly dominated by their valuable *antimalarial* and *oral hypoglycemic* properties. These functions have been closely studied in an effort to gain an understanding of the mode of action of these compounds. The present review, dealing primarily with the chemistry of biguanides, cannot attempt to provide more than the briefest outline of the voluminous literature concerned with the medical uses of biguanides, but an effort has been made to list all relevant references given in Chemical Abstracts.

### 1. Antimalarial Properties

Several excellent accounts have described the need which arose in World War II for new antimalarial drugs, and the chemical reasoning which led to the synthesis of Paludrine (1-p-chlorophenyl-5-isopropylbiguanide) (*142, 146, 545*). *Curd, Davey* and *Rose* (*142*) found that both Paludrine and its 5-methyl-homologue showed antimalarial activity in chicks and birds; clinical trials demonstrated their activity in man. Paludrine presently proved (*2*) to be of considerable therapeutic value in the prevention and cure of acute human malaria (*415*). Animal experiments (ducks) indicated (*178, 421*) that Paludrine was most effectively administered in the diet in order to maintain a suitable concentration in the blood.

In view of the remarkable activity and non-toxicity of Paludrine as an antimalarial, surpassing the previously used drugs quinine, Mepacrine and Pamaquine, its mode of action and pharmacology attracted considerable interest. Immediately following its discovery, studies concerning its toxicity (*89, 103*), absorption, distribution and excretion in animals (*651*) and man (*2*) and those of its 5-methyl homologue (*654*) were carried out. It was soon recognised (*65, 751*) and confirmed (*63, 66*) that malarial parasites are able to acquire resistance (*64, 340*) to both Paludrine and its 5-methyl analogue.

In studies aiming to elucidate the mode of action of this drug, a significant observation (*67*) was the fact that Paludrine and its 5-methyl homologue showed antimalarial *in vivo* activity (*67, 716*), but that both these compounds were inactive against malaria-infected cells *in vitro* (*716*). This suggested that Paludrine itself is inactive, but that it is converted into an active metabolite in the body. Efforts (*422*) to ascertain the mode of action of Paludrine led to *Crowther* and *Levi's* (*136*) isolation of its active metabolite which proved to be 4,6-diamino-1-p-chlorophenyl-1,2-dihydro-2,2-dimethyl-1,3,5-triazine; a detailed account of the purely chemical work which led to the assignment of this structure is available (*95*). The metabolites, though highly active, are less useful antimalarials than their parent compounds (*541*) because of their rapid excretion from the organism.

The mode of action (*561*) of both Paludrine and its active metabolite, and their interconversion (*627*) have been studied, as well as those of several of their analogues (*33, 72, 232, 526, 540, 561, 621*).

A method of estimating the degree of protection from malaria in man has been described (*168*) and most aspects of malaria, including the chemotherapeutic action of Paludrine, have been reviewed (*70, 233, 285, 398*).

In view of the exceptional medical interest of Paludrine, the range of studies on its pharmacological and physiological behaviour has been very extensive (*4, 18, 21, 71, 103, 114, 169, 170, 215, 329, 330, 341, 348, 431, 507, 523, 575, 592, 620, 658, 695, 755, 761*).

It has been claimed (*635*) that the copper complex of Paludrine has antirheumatic and anti-inflammatory properties, and that Paludrine possesses (*431*) diuretic activity.

## 2. Oral Hypoglycemic Properties

As early as 1929 biguanides had been recognised by *Slotta* and *Tschesche* (*626*) to possess oral hypoglycemic properties. Later investigators discussed the use of alkylenebiguanides as insulin-substitutes (*284*), reported a slight hypoglycemic activity of Paludrine (*103*) and 1,1-dimethylbiguanide hydrochloride (*232*) and examined the toxicity and glycemic action of aryl- and alkylbiguanides (*732*). However, none of these compounds proved successful in the treatment of diabetes.

The first product useful in this respect, reported by *Ungar, Freedman* and *Shapiro* (*726*), was β-phenethylbiguanide, which proved highly active as an oral hypoglycemic agent for many species and showed a curative action in human diabetes. A clinical evaluation (*364*) of its antidiabetic properties followed rapidly. The pharmacology of the drug has been studied in great detail (*23, 29, 39, 49, 55, 69, 90, 124, 182, 328, 366, 425, 434, 472, 483, 492, 494, 495, 499, 506, 514, 623, 640, 641, 659, 665, 678, 726, 733, 734, 738, 748—750, 756*) and its toxicity measured (*500*). Reviews on its use as an antidiabetic drug are available (*125, 197, 191, 363, 364, 429, 430, 715*). A selection of references dealing with biological and physiological aspects of the behaviour of β-phenethylbiguanide are given in Table 10.

Table 10. *Biological and Physiological studies with β-phenethylbiguanide*

| Studies referring to | References |
|---|---|
| Carbohydrate Metabolism | (*55, 283, 350, 402, 468, 469, 470, 520, 692, 722, 727, 744, 747*) |
| Protein Metabolism | (*289, 402, 745*) |
| Fat Metabolism | (*80, 182, 183, 396*) |
| Steroid Metabolism | (*47, 164, 406, 474*) |
| Clinical Aspects | (*11, 165, 337, 639*) |
| Side Effects | (*159, 381, 499, 678*) |
| Other Aspects | (*1, 77, 205, 280, 326, 366, 413, 450, 551, 611, 663*) |

*Other Hypoglycemic Biguanides*

Amongst other biguanides of potential antidiabetic interest, 1-butylbiguanide and 1,1-dimethylbiguanide are the most important.

1-*Butylbiguanide*. The hypoglycemic activities of biguanide-homologues vary from species to species (*388, 640*). The usefulness of butylbiguanide as an antidiabetic agent is well-established, and numerous pharmacological studies have dealt with this and related properties (*123, 124, 174, 334, 388, 389, 390, 480, 481, 514, 640, 641*).

High dosage (*514*) of butylbiguanide, even seventy times (*279*) the therapeutic dose, did not produce any pathological changes in animals.

1,1-*Dimethylbiguanide*. Numerous studies have demonstrated the antidiabetic properties of 1,1-dimethylbiguanide (*28, 76, 232, 641, 660*) and its pharmacology has attracted much attention (*28, 76, 116, 196, 232, 479, 480, 641, 660, 674, 719*). A favourable report on five years' experience in the use of this drug has appeared (*662*).

*Other Examples*. Good antidiabetic properties have been claimed (*594*) for compounds of the type $RR'N \cdot C(:NH)NHC(:NH)NH_2$ where

(a) R = Alkyl of four or five carbon atoms; R' = H

(b) R = $Aryl(CH_2)_n \cdot$ (n = 1 or 2); R' = H

(c) R = Benzyl; R' = Methyl.

This group of compounds is further reported (*595*) to be helpful in the treatment of mental diseases, especially schizophrenia. Other biguanides that have been examined for their antidiabetic potency are benzylbiguanide (*334, 494, 495*), α-phenethylbiguanide (*494, 495*) and amyl- and isoamyl-biguanide (*749*). The hypoglycemic activity and toxicity of a wide range of biguanides has been recorded (*496, 604*) and an attempt made to correlate their structures and activity (*496*). The effectiveness of copper complexes of biguanides as hypoglycemic agents has been considered (*506*).

## 3. Tumor-Inhibiting Properties

Phenanthrenebiguanides possess (*384*) some activity against cancer. The effect of biguanides on plasma protein surface has been investigated (*294, 610*) with a view to the possibility of finding anti-tumor agents. Aryl- and naphthyl-biguanides exhibit (*625*) a slight anti-neoplastic activity *in vitro* in Ehrlich ascites carcinoma.

## 4. Antibacterial Action

An exceptionally wide range of biguanides are reported to be bactericides. In one investigation 242 mono- to tetra-substituted biguanides were tested *in vitro* (*509*) for antibacterial and antimycotic activity, and over 50% were effective. On the other hand, only *one* of a large number of arylbiguanides tested for bactericidal action was found (*224*) to be markedly active. Bactericidal biguanide derivatives are listed in Table 11. The action of a number of biguanides on Skigella flexnem bacteriophages has been investigated (*387*).

## 5. Tuberculostatic Action

4-Amino-4'-biguanidodiphenyl sulphone and 4,4'-diguanidophenyl sulphone have no antitubercular activity (*736*), but the hydrochlorides of 2-naphthylbiguanide (*731*), (p-nitrophenyl)biguanide (*731*) and 8-quinolylbiguanide (*731*) have some antitubercular activity *in vitro*, as have biguanidodiphenyl (*160*) and derivatives of biguanidodiphenyl ether (*333*).

## 6. Anti-Virial Properties

1,1-Dimethylbiguanide has been said (*232*) to be analgesic, antipyretic, non-toxic and possibly of value as an anti-influenza drug, but its antipyretic properties have not been confirmed (*675*) The compound inhibits the course of cowpox, influenza A, pneumonia and measles (*676*).

Table 11. *Bactericidal biguanides*

| Compound | Reference |
|---|---|
| 1-Acridyl-5-arylbiguanide | (*154*) |
| Alkoxybiguanides | (*418, 512*) |
| Aryl(alk)oxybiguanides | (*418*) |
| p-Biguanidoacetophenone amidinohydrazone and analogues | (*108*) |
| Biguanidodiphenyl ether derivatives | (*333*) |
| Biguanido-substituted Diphenylene Oxides | (*515*) |
| 1-p-Chlorophenyl-5-methyl-5-isopropylbiguanide | (*224*) |
| 1,5-Bis(3,4-dichlorobenzyl)biguanide hydrochloride | (*316, 741*) |
| 1-p-Chlorophenyl-5-phenylsulphonamido-2-thiazole-biguanide hydrochloride | (*463*) |
| 1-p-Chlorophenyl-5-p-sulphonamidophenylbiguanide | (*463*) |
| 1,5-Dibenzylbiguanides | (*231*) |
| Diphenylbiguanides | (*107*) |
| Dinitrophenol salts of Arylbiguanides | (*309*) |
| Lauroxylpropylbiguanides | (*219, 220*) |
| Long-chain Aliphatic Biguanides | (*516*) |

Table 11 (continued)

| Compound | Reference |
|---|---|
| Methoxyphenylbiguanides | (*646*) |
| Paludrine acetate | (*220*) |
| β-Phenethylbiguanide | (*342, 375, 522, 680, 742*) |
| Piperazinylbiguanides | (*339*) |
| 5-Quinolylbiguanides | (*622*) |

The salts of heterylbiguanides are generally active (*8*) against viruses, and are of unusually low toxicity. They appear to be useful against influenza, herpes zoster, smallpox, Newcastle disease and canine distemper. Xylylbiguanides are also active (*112*) against influenza, and isopropylbiguanide hydrochloride inhibits Lee influenza virus in chick embryos (*740*).

1,1-Anhydrobis(β-hydroxyethyl)biguanide hydrochloride is an influenza suppressant in mice (*432*). Butylbiguanide and hexylbiguanide have some inhibitory action on viruses (*226*), as have biguanidodiphenyl ether derivatives (*333*). The anti-virial activity of some additional aryl- and alkylbiguanides has been examined *in vitro* (*334*).

## B. Dibiguanides

Dibiguanides are active against trypanosomiasis (*318*). Dibiguanides of the type

$$\text{Aryl-NH}\cdot\underset{\underset{\text{NH}}{\|}}{\text{C}}\cdot\text{NH}\cdot\underset{\underset{\text{NH}}{\|}}{\text{C}}\cdot\text{NH}\cdot(\text{CH}_2)_n\cdot\text{NH}\cdot\underset{\underset{\text{NH}}{\|}}{\text{C}}\cdot\text{NH}\cdot\underset{\underset{\text{NH}}{\|}}{\text{C}}\cdot\text{NH-Aryl}$$

(n=3 to 9) are bactericides (*550*), having maximum activity when n=5 to 7 (*549, 550*).

A wide range of dibiguanides have been synthesised (*107, 157, 587*) as bactericides, fungicides and sterilising agents (*550*). The most effective of these is (*171*) 1,6-di-p-chlorophenylbiguanidohexane ("Hibitan"), having a high bactericidal activity (*22, 171, 414, 573*) and wide potential applications (*312, 324, 351, 401, 438, 459, 588, 607*).

# X. Additional Physiological Properties

The greater part of the researches dealing with physiological and pharmacological aspects of biguanides have been cited in Section IX dealing with their medical uses: this discussion was therefore concerned mostly with Paludrine, β-phenethylbiguanide, and related compounds. Physiological, pharmacological and allied investigations on additional biguanide derivatives are tabulated in the present Section (see Table 12).

Table 12. *Physiological and pharmacological properties of biguanides*

| Compound | Reference |
|---|---|
| Biguanide | (*505, 657, 720*) |
| Biguanide, Penicillin Salt | (*739*) |
| Phenylbiguanide | (*40, 88, 176, 186, 223, 319, 407, 431 460, 490, 491, 737*) |
| o-Tolylbiguanide | (*96, 311, 577*) |
| p-Chlorophenylbiguanide | (*229*) |
| Arylbiguanides | (*175, 538, 591*) |
| 1-Aryl-5-alkylbiguanides | (*175*) |
| Quinazolylbiguanide (labelled) | (*115*) |
| Miscellaneous biguanides | (*542, 553, 661, 746*) |

## XI. Industrial and other Uses

A large number of publications, many of them in the patent literature, have described actual and potential uses of biguanides for a variety of purposes. They are collected in Table 13, and are classified according to the area of usefulness to which they refer.

Table 13. *Industrial uses of biguanides*

| Application | Reference |
|---|---|
| Adhesives | (*636*) |
| Anticorrosion Agents | (*104, 118, 317, 331, 691*) |
| Antistaining Agents | (*315*) |
| Antistatic Agents | (*93, 94, 451, 452, 579*) |
| Detergents | (*94, 270, 752*) |
| Dyeing, Use of Biguanides | (*98, 199, 255, 309, 367, 393, 394, 395, 409*) |
| Dyeing, Use of Biguanide-Condensation Products | (*50, 75, 82, 83, 207, 209, 239, 240, 399, 410, 504, 564, 565, 566, 567, 568, 637, 667, 692*) |
| Dyeing, Use of Acylbiguanides and Sulphonated Biguanides | (*13, 53, 54, 172, 234, 235, 238, 243, 244, 245, 247, 251, 253, 254, 257, 268, 275*) |
| Floating Agents (Ore-Dressing) | (*13, 172*) |
| Hair-Dressing | (*762*) |
| Insecticides and Fungicides | (*9, 62, 309, 550, 628, 763*) |

Table 13 (continued)

| Application | Reference |
|---|---|
| Intermediates | (*346, 707*) |
| Ion Exchange Resins | (*12, 344, 679, 718*) |
| Leather and Tanning | (*290, 447, 458, 462, 717*) |
| Moth Repellents | (*448, 550*) |
| Paper | (*290, 320, 321*) |
| Photography | (*102, 105, 179, 189, 192, 349, 433, 560, 579, 624, 655*) |
| Plant Disinfectants | (*261, 273, 274, 550*) |
| Plastics | (*81, 84, 162, 163, 172, 713*) |
| Preservatives | (*571*) |
| Resins | (*20, 214, 216, 447, 619*) |
| Rubber | (*87, 359, 382, 447, 499, 543, 559, 725*) |
| Soldering Flux | (*503*) |
| Solubilising Agent | (*386*) |
| Stabilisers and Antoxidants: | |
| a) Fats and Soaps | (*119, 120, 121, 210, 358, 760*) |
| b) Polymers | (*180, 310*) |
| c) Other Compounds | (*92, 187, 299, 308, 358, 403, 404, 405, 539, 570, 656*) |
| Surface Active Agents | (*53, 54, 234, 235, 246, 257, 258, 259, 260, 266, 269, 270, 271, 272, 276, 516, 637*) |
| Textiles: | |
| a) Biguanides | (*54, 181, 198, 230, 241, 242, 248, 249, 250, 277, 290, 320, 385, 664, 743*) |
| b) Condensation Products of Biguanides | (*252, 256, 260, 262, 263, 264, 265, 267, 270, 275, 286, 400, 448, 478, 637*) |

# XII. References


1. *Acharyya, B. P., U. N. De, S. K. Mukherjee, M. A. Hai,* and *H. M. Chakravarty:* Indian J. Med. Res. *49,* 62 (1961).
2. *Adams, A. R. D., B. G. Maegraith, J. D. King, R. H. Townsend, T. H. Davey,* and *R. E. Havard:* Ann. Trop. Med. Parasitol. *39,* 225 (1945).
3. –, *R. H. Townsend,* and *J. D. King:* Ann. Trop. Med. Parasitol. *39,* 217 (1945).
4. *Agarwal, S. L.,* et *B. S. Deshmanker:* Arch. Intern. Pharmacodyn. *123,* 305 (1960).
5. *Ainley, A. D., F. H. S. Curd,* and *F. L. Rose:* J. Chem. Soc. *1949,* 98.
6. — — — Brit. Patent 607,720 (1948); Chem. Abstracts *43,* 4691 (1949).
7. — — — Brit. Patent 624,578 (1949); Chem. Abstracts *44,* 3014 (1950).
8. *Aktiebolaget Kabi Ltd.:* Brit. Patent 776,176 (1957); Chem. Abstracts *51,* 16569 (1957).
9. *Albert, A.,* and *D. Magrath:* Biochem. J. *41,* 534 (1947).
10. *Alekseeva, L. V.,* and *Z. V. Pushkareva:* Zh. Obshch. Khim. *35,* 1693 (1963).
11. *Alvarez, L. M., I. Ganopol, R. E. Pupi* y *J. F. Rofaele:* Semana Med. (Buenos Aires) *122,* 432 (1963); Chem. Abstracts *59,* 14485 (1963).
12. *American Cyanamid Co.:* Brit. Patent 562,402 (1944); Chem. Abstracts *40,* 660 (1946).
13. – Brit. Patent 573,310 (1945); Chem. Abstracts *43,* 5726 (1949).
14. – Brit. Patent 631,878 (1949); Chem. Abstracts *44,* 4027 (1950).
15. – Brit. Patent 643,012 (1950); Chem. Abstracts *45,* 5180 (1951).
16. *Andersag, H.,* and *K. Westphal:* Ber. *70,* 2035 (1937).
17. *Andreasch, R.:* Monatsh. Chem. *48,* 145 (1927).
18. *Angelakos, E. T.,* and *A. H. Hegnauer:* J. Pharmacol. Exptl. Therap. *127,* 137 (1959).
19. *Angelucci, R., D. Artini, P. N. Giraldi, W. Logemann* i *G. Nannini:* Farmaco, Ed. Sci. *16,* 633 (1961).
20. *Appelquest, A. J.:* U. S. Patent 2,517,824 (1950); Chem. Abstracts *44,* 10,375 (1950).
21. *Arora, R. B., P. L. Sharma, U. N. Gupta, A. Lal,* and *C. N. Mathur:* Arch. Intern. Pharmacodyn. *124,* 386 (1960).
22. *Ashikari, Y., S. Sakuma, H. Hasegawa,* and *M. Tsuruoka:* Tanabe Seiyaku Kenkyu Nempo *5,* 34 (1960); Chem. Abstracts *55,* 9663 (1961).
23. *Ashkar, E., C. N. Burrier,* and *M. C.* de *P. Ramos:* Rev. Soc. Arg. Biol. *34,* 11 (1958).
24. *Ashworth, R.* de *B., A. F. Crowther, F. H. S. Curd, J. A. Hendry, D. N. Richardson,* and *F. L. Rose:* J. Chem. Soc. *1949,* 475.
25. –, *F. H. S. Curd, J. A. Hendry,* and *F. L. Rose:* Brit. Patent 599,714 (1948); Chem. Abstracts *42,* 7790 (1948).
26. *Bailey, R. E.,* and *D. A. Durfee:* J. Chromatog. *16,* 546 (1964).
27. *Baker, B. R.,* and *B. T. Ho:* J. Heterocyclic Chem. *2,* 72 (1965).
28. *Balasse, E.:* Compt. Rend. Soc. Biol. *153,* 1892 (1959).

29. —, et *V. Conrad:* Rev. Franc. Etudes Clin. Biol. *7,* 803 (1961).
30. *Bamberger, E.:* Ber. *13,* 1581 (1880).
31. —, and *W. Dieckmann:* Ber. *25,* 545 (1892).
32. *Bami, H. L.:* Current Science (India) *18, 77* (1949).
33. — Indian J. Malariol. *7,* 283 (1953).
34. — J. Sci. Ind. Res. *14C,* 231 (1955).
35. —, and *P. C. Guha:* Sci. Cult. (Calcutta) *14,* 386 (1949).
36. —, *B. M. Iyer,* and *P. C. Guha:* J. Indian Inst. Sci. *29A,* 1 (1946).
37. *Bandyapadhaya, D., N. N. Gosh,* and *P. Ray:* J. Indian Chem. Soc. *29,* 157 (1952).
38. *Banerjea, D.,* and *B. Chakravarty:* J. Inorg. Nucl. Chem. *26,* 1233 (1964).
39. *Bangham, A. D., J. C. Glover, S. Hollingshead,* and *B. A. Pethica:* Biochem. J. *84,* 513 (1962).
40. *Barer, G. R.,* and *E. Nuesser:* Brit. J. Pharmacol. *13,* 372 (1958).
41. *Basu, U. P., K. R. Chandran,* and *A. K. Sen:* J. Indian Chem. Soc. *28,* 467 (1951).
42. *Basu, A. R., S. Gupta,* and *A. N. Bose:* J. Sci. Ind. Res. (India) *9B,* 57 (1950).
43. —, *A. K. Sen,* and *A. K. Ganguly:* Sci. Cult. (Calcutta) *18,* 45 (1952).
44. *Bauer, L., J. Cymerman,* and *W. J. Sheldon:* J. Chem. Soc. *1951,* 2342.
45. *Bekhli, A. F., V. N. Uyimtsev,* and *K. S. Topchiev:* Zh. Prikl. Khim. *20,* 591 (1947); Chem. Abstracts *43,* 3793 (1949).
46. *Beral, H., V. Stoicescu,* and *C. Ivan:* Farmacia (Bucharest) *12,* 631 (1964); Chem. Abstracts *62,* 3409 (1965).
47. *Bergen, S. S., J. G. Hilton,* and *W. S. Norton:* Proc. Soc. Exptl. Biol. Med. *98,* 625 (1958).
48. *Bernshtein, V. N.,* and *I. V. Chuiko:* Uch. Zap. Pyatigorsk Gos. Farmatsevt. Inst. *5,* 193 (1961); Chem. Abstracts *59,* 1440 (1963).
49. *Bertarelli, P.:* Boll. Chim. Farm. *97,* 396 (1958).
50. *Bessot, L. E.,* and *P. Garet:* U. S. Patent 2,556,315 (1951); Chem. Abstracts *46,* 1039 (1952).
51. *Beyer, H., H. Bieling,* and *T. Pyl:* Z. Chem. *2,* 310 (1962).
52. *Bhattacharya, B., A. K. Acharya,* and *U. P. Basu:* Indian J. Chem. *2,* 370 (1964).
53. *Bindler, J.:* U. S. Patent 2,451,432; Chem. Abstracts *43,* 699 (1949).
54. —, and *H. Schlaepfer:* U. S. Patent 2,324,354 (1943).
55. *Biro, L., T. Banyasz, M. B. Kovacs* u. *M. Bajor:* Klin. Wochschr. *39,* 760 (1961).
56. *Birtwell, S.:* J. Chem. Soc. *1952,* 1279.
57. —, *A. F. Crowther, F. H. S. Curd, J. A. Hendry, D. N. Richardson,* and *F. L. Rose:* Brit. Patent 603,069 (1948); Chem. Abstracts *43,* 683 (1949).
58. —, *F. H. S. Curd, J. A. Hendry,* and *F. L. Rose:* J. Chem. Soc. *1948,* 1645.
59. — —, and *F. L. Rose:* J. Chem. Soc. *1949,* 2556.
60. — — — Brit. Patent 637,892 (1950); Chem. Abstracts *44,* 9982 (1950).
61. — — — Brit. Patent 638,695 (1950); Chem. Abstracts *44,* 9983 (1950).
62. —, and *F. L. Rose:* Brit. Patent 785,937 (1957); Chem. Abstracts *52,* 11921 (1958).
63. *Bishop, A.:* Intern. Congr. Biochem., Abstr. of Communs., First Congr., Cambridge *1949,* 437; Chem. Abstracts *47,* 203 (1953).
64. — Parasitology *52,* 495 (1962).
65. —, and *B. Birkett:* Nature *159,* 884 (1947).
66. — — Parasitology *39,* 125 (1948).

67. *Black, R. H.:* Med. Colonial *9,* 259 (1947); Trans. Roy. Soc. Trop. Med. Hyg. *40,* No. 2 (1946).
68. *Blair, J. S.,* and *J. M. Brabham:* J. Am. Chem. Soc. *44,* 2346 (1922).
69. *Blakely, R. M.,* and *H. I. Macgregor:* Can. J. Animal Sci. *42,* 102 (1962).
70. *Blanchard, K. C.:* Ann. Rev. Biochem. *16,* 587 (1947).
71. *Blaschko, H., T. C. Chou,* and *I. Wajda:* Brit. J. Pharmacol. *2,* 116 (1947).
72. *Bliznyukov, V. I.:* Farmatsevt. Zh. (Kiev) *14,* No. 5, 6 (1959); Chem. Abstracts *55,* 25814 (1961).
73. —, and *L. S. Sokol:* Farmatsevt. Zh. (Kiev) *15,* 12 (1960); Chem. Abstracts *55,* 13429 (1961).
74. — —, and *N. T. Solonskaya:* Zh. Obshch. Khim. *34,* 329 (1964).
75. *Boeckmann, K., K. Taube,* and *O. Weber:* German Patent 896,439 (1953); Chem. Abstracts *52,* 13279 (1958).
76. *Boeringer, C. F.:* Brit. Patent 842,925 (1960); Chem. Abstracts *55,* 8300 (1961).
77. *Bose, A. N., S. P. Paul,* and *U. P. Basu:* Sci. Cult. (Calcutta) *26,* 86 (1960).
78. *Bourdais, J.:* Bull. Soc. Chim. France *1962,* 1174.
79. *Bradley, M. H.:* U. S. Patent 2,309,624 (1943).
80. *Brand, W. von:* Arzneimittel-Forsch. *11,* 739 (1961).
81. *British Thomson-Houston Co. Ltd.:* Brit. Patent 558,051 (1943); Chem. Abstracts *40,* 241 (1946).
82. *Brodersen, K.:* German (East) Patent 10,050 (1955); Chem. Abstracts *52,* 19154 (1958).
83. — German (East) Patent 11,019 (1956); Chem. Abstracts *53,* 6634 (1959).
84. —, and *M. Quaedvlieg:* German Patent 737,543 (1943); Chem. Abstracts *39,* 5128 (1945).
85. *Brookes, A.,* (to *American Cyanamid Co.*); U. S. Patent 2,287,597 (1943); Chem. Abstracts *37,* 144 (1943).
86. *Brown, C. J.:* J. Chem. Soc. (A), *1967,* 60.
87. *Bruers, W., H. Wolff,* and *E. Ranft:* German (East) Patent 12027 (1956); Chem. Abstracts *53,* 757 (1959).
88. *Buellbring, E.:* Brit. J. Pharmacol. *13,* 444 (1958).
89. *Butler, R., D. G. Davey,* and *A. Spinks:* Brit. J. Pharmacol. *2,* 181 (1947).
90. *Butterfield, J., I. Kelsey,* and *E. Holling:* Diabetes *7,* 449 (1958).
91. *Bymerman, H. C., J. S. Buntekoe,* and *W. L. C. van der Berg:* Rec. Trav. Chim. *73,* 109 (1954).
92. *Cabal, A. V.,* and *H. G. Oliver:* U. S. Patent 2,463,478 (1949); Chem. Abstracts *43,* 4701 (1949).
93. *Carnes, J. J.,* and *W. T. Booth jun.:* U. S. Patent 2,652,348 (1953); Chem. Abstracts *48,* 1060 (1954).
94. —, *J. A. Price,* and *W. T. Booth jun.:* U. S. Patent 2,739,168 (1956); Chem. Abstracts *50,* 9756 (1956).
95. *Carrington, H. C., A. F. Crowther,* and *G. J. Stacey:* J. Chem. Soc. *1954,* 1017.
96. *Catsch, A.,* and *Le Du-Khuong:* Nature *180,* 609 (1957).
97. *Cattapan, D.,* i *A. Vercellone:* Gazz. Chim. Ital. *85,* 345 (1955).
98. *Cellarius, L.:* German Patent 803,830 (1951); Chem. Abstracts *46,* 10634 (1952).
99. *Chakravarty, K.,* and *P. Ray:* J. Indian Chem. Soc. *21,* 41 (1944).
100. *Chatterjee, A. K.:* Sci. Cult. (Calcutta) *20,* 46 (1954).
101. — Sci. Cult. (Calcutta) *20,* 246 (1954).
102. *Chechak, J. J.:* U. S. Patent 2,418,623 (1947); Chem. Abstracts *42,* 48 (1948).
103. *Chen, K. K.,* and *R. C. Anderson:* J. Pharmacol. Exptl. Therap. *91,* 157 (1947).

104. *Chenicek, J. A.*, and *R. B. Thompson:* U. S. Patent 2,734,807 (1956); Chem. Abstracts *50,* 8437 (1956).
105. *Ciba Ltd.:* Belgian Patent 622,239 (1963); Chem. Abstracts *59,* 1220 (1963).
106. — Belgian Patent 630,463 (1963); Chem. Abstracts *61,* 4273 (1964).
107. — Brit. Patent 890,477 (1962); Chem. Abstracts *57,* 3363 (1962).
108. — Brit. Patent 894,809 (1962); Chem. Abstracts *57,* 9742 (1962).
109. — Swiss Patent 209,503 (1940); Brit. Patent 527,697 (1940); French Patent 849,752 (1940); German Patent 715,761 (1941).
110. *Cincotta, J. J.*, and *R. Feinland:* Anal. Chem. *34,* 774 (1962).
111. *Clark, E. G. C.:* J. Pharm. Pharmacol. *10,* 194 (1958).
112. *Clark, R. J., A. Isaacs,* and *J. Walker:* Brit. J. Pharmacol. *13,* 424 (1958).
113. *Clauder, O., B. Zemplen,* and *G. Bulscu:* Austrian Patent 168,063; Chem. Abstracts *47,* 8097 (1953).
114. *Coatney, G. R.*, and *W. C. Cooper:* J. Parasitol. *34,* 275 (1948).
115. *Cohen, Y.:* Prod. Pharm. *16,* 341 (1961); Chem. Abstracts *56,* 884 (1962).
116. —, and *O. Costerousse:* Congr. Federation Intern. Diabete, 4e, Geneva, Switz. *1,* 745 (1961); Chem. Abstracts *59,* 15779 (1963).
117. *Cohn, G.:* J. Prakt. Chem. *84,* 396 (1911).
118. *Compagnies Reunies* des huileries du Congo Belge et Savonneries Lever Freres „Huilever" Soc. anon., Belg. Patent 510,621 (1952); Chem. Abstracts *52,* 10616 (1958).
119. *Cook, E. W.:* Brit. Patent 591,836 (1947); Chem. Abstracts *42,* 3978 (1948).
120. — U. S. Patent 2,375,626 (1945); Chem. Abstracts *39,* 3452 (1945).
121. — U. S. Patent 2,467,295 (1949); Chem. Abstracts *43,* 5215 (1949).
122. *Cramer, W.:* Ber. *34,* 2594 (1910).
123. *Creutzfeldt, W., U. Deuticke,* u. *H. D. Soeling:* Klin. Wochschr. *39,* 790 (1961).
124. —, *H. D. Soeling, A. Moench, E. Rauh* u. *M. Bol:* Arch. Exptl. Pathol. Pharmakol. *244,* 31 (1962).
125. — —, and *Z. Zarday:* Metab. Clin. Exptl. *12,* 264 (1963).
126. *Crounse, N. N.:* J. Org. Chem. *16,* 492 (1951).
127. *Crowther, A. F.:* Brit. Patent 776,679; Chem. Abstracts *52,* 1289 (1958).
128. —, and *F. H. S. Curd:* Brit. Patent 649,372 (1951); Chem. Abstracts *46,* 1038 (1952).
129. — — Brit. Patent 649,692 (1951); Chem. Abstracts *46,* 1038 (1952).
130. — — Brit. Patent 649,693 (1951); Chem. Abstracts *46,* 1039 (1952).
131. — — Brit. Patent 667,116 (1952); Chem. Abstracts *47,* 5435 (1953).
132. — —, *D. G. Davey, J. A. Hendry, W. Hepworth,* and *F. L. Rose:* J. Chem. Soc. *1951,* 1774.
133. — —, *D. N. Richardson,* and *F. L. Rose:* J. Chem. Soc. *1948,* 1636.
134. — —, and *F. L. Rose:* J. Chem. Soc. *1951,* 1780.
135. — — — Brit. Patent 618,613 (1949); Chem. Abstracts *43,* 6227 (1949).
136. —, and *A. A. Levi:* Brit. J. Pharmacol. *8,* 93 (1953).
137. *Cuiko, I. V.*, and *V. N. Bernshtein:* Peredovyl Metody Khim. sb. 1964, 264; Chem. Abstracts *63,* 8124 (1964).
138. *Curd, F. H. S., N. W. Cusa,* and *A. G. Murray:* Brit. Patent 667,094 (1952); Chem. Abstracts *47,* 5433 (1953).
139. — —, and *C. H. Vasey:* Brit. Patent 662,467 (1951); Chem. Abstracts *46,* 11238 (1952).
140. —, *D. G. Davey,* and *R. de B. Ashworth:* J. Chem. Soc. *1949,* 1732.
141. — —, *D. N. Richardson,* and *R. de B. Ashworth:* J. Chem. Soc. *1949,* 1739.
142. — —, and *F. L. Rose:* Ann. Trop. Med. Parasitol. *39,* 208 (1945).

143. —, *J. A. Hendry, T. S. Kenny, A. G. Murray,* and *F. L. Rose:* J. Chem. Soc. *1948,* 1630.
144. —, and *D. N. Richardson:* U. S. Patent 2,544,827 (1951); Chem. Abstracts *45,* 8041 (1951).
145. — —, and *F. L. Rose:* Brit. Patent 619,498 (1949); Chem. Abstracts *43,* 6658 (1949).
146. —, and *F. L. Rose:* Chem. Ind. (London) *1946,* 74.
147. — — J. Chem. Soc. *1946,* 362.
148. — — J. Chem. Soc. *1946,* 729.
149. — — Brit. Patent 577,843 (1946); Chem. Abstracts *41,* 3121 (1947).
150. — — Brit. Patent 581,346 (1946); Chem. Abstracts *41,* 3126 (1947).
151. — — U. S. Patent 2,467,371 (1949); Chem. Abstracts *43,* 6659 (1949).
152. — — U. S. Patent 2,510,081 (1950); Chem. Abstracts *44,* 8366 (1950).
153. — — U. S. Patent 2,531,404 (1950); Chem. Abstracts *45,* 4265 (1951).
154. — — U. S. Patent 2,531,405 (1950); Chem. Abstracts *45,* 4266 (1951).
155. *Cuthbertson, W. W.,* and *J. S. Moffat:* J. Chem. Soc. *1948,* 561.
156. *Cutler, R. A.:* U. S. Patent 2,836,539 (1958); Chem. Abstracts *52,* 20218 (1958).
157. —, and *S. Schalit:* French Patent M1582 (1962); Chem. Abstracts *59,* 6312 (1963).
158. — — U. S. Patent 3,136,816 (1964); Chem. Abstracts *61,* 9512 (1964).
159. *Cutting, W.:* Antibiot. Chemotherapy *12,* 671 (1962).
160. *Cymerman-Craig, J., S. D. Rubbo,* and *B. J. Pierson:* Brit. J. Exptl. Pathol. *36,* 254 (1955).
161. *Daebritz, E.:* Angew. Chem. *78,* 483 (1966); Internat. Ed. *5,* 470 (1966).
162. *D'Alelio, G. F.:* U. S. Patent 2,331,376 (1943); Chem. Abstracts *38,* 1534 (1944).
163. — U. S. Patent 2,331,377 (1943); Chem. Abstracts *38,* 1534 (1944).
164. *Dalidowicz, J. E.,* u. *H. J. McDonald:* Naturwissenschaften *49,* 422 (1962).
165. *Danowski, T. S.,* and *F. M. Mateer:* Proc. Soc. Exptl. Biol. Med. *102,* 639 (1959).
166. *Dansi, A.,* i *C. Zanini:* Gazz. Chim. Ital. *89,* 1681 (1959).
167. — — Boll. Chim. Farm. *98,* 580 (1959); Chem. Abstracts *54,* 6597 (1960).
168. *Davey, D. G.,* and *G. I. Robertson:* Trans. Roy. Soc. Trop. Med. Hyg. *51,* 450 (1957).
169. — — Trans. Roy. Soc. Trop. Med. Hyg. *51,* 463 (1957).
170. — — Trans. Roy. Soc. Trop. Med. Hyg. *51,* 502 (1957).
171. *Davies, G. E., J. Francis, A. R. Martin, F. L. Rose,* and *G. Swain:* Brit. J. Pharmacol. *9,* 192 (1954).
172. *Davis, A. R.:* U. S. Patent 2,389,718 (1945); Chem. Abstracts *40,* 1880 (1946).
173. *Davis, T. L.:* J. Am. Chem. Soc. *43,* 2234 (1921).
174. *Daweke, H.,* and *I. Bach:* Metab. Clin. Exptl. *12,* 319 (1963).
175. *Dawes, G. S.,* and *J. C. Mott:* Brit. J. Pharmacol. *5,* 65 (1950).
176. — —, and *J. G. Widelicome:* Arch. Intern. Pharmacodyn. *90,* 203 (1952).
177. *De, A. K., N. Gosh,* and *P. Ray:* J. Indian Chem. Soc. *27,* 493 (1950).
178. *Dearborn, E. H.,* and *E. K. Marshall jun.:* Proc. Soc. Exptl. Biol. Med. *63,* 46 (1946).
179. *Delangre, J. P.:* U. S. Patent 2,821,455 (1958); Chem. Abstracts *52,* 6033 (1958).
180. *Depree, D. O.,* and *E. F. Hill:* U. S. Patent 2,729,614 (1956); Chem. Abstracts *50,* 11050 (1956).
181. *Dithmar, K.,* and *E. Naujoks:* German Patent 1,086,664: Chem. Abstracts *55,* 26469 (1961).
182. *Ditschuneit, H., W. Lotz, W. Fritzsche,* and *E. F. Pfeiffer:* Cong. Federation Intern. Diabete, *4,* Geneva, Switz. 1961, *1,* 740; Chem. Abstracts *57,* 13131 (1962).

183. —, *E. F. Pfeiffer, E. Blay, H. G. Rossenbeck* u. *K. Schoeffling*, Symp. Deut. Ges. Endokrinol. *7*, 194 (1960).
184. *Dittler, E.:* Monatsh. *29*, 645 (1908).
185. *Doub, L.:* U. S. Patent 3,170,925 (1965); Chem. Abstracts *63*, 1808 (1965).
186. *Douglas, W. W.*, and *J. M. Ritchie:* J. Physiol. *138*, 31 (1957).
187. *Downing, F. B.*, and *C. J. Pedersen:* U. S. Patent 2,373,021 (1945); Chem. Abstracts *39*, 3424 (1945).
188. *Drake, N. L.*, and *R. J. Kray:* J. Am. Chem. Soc. *76*, 1320 (1954).
189. *Dreyfus, P. D.:* U.S. Patent 2,368,647 (1945); Chem. Abstracts *39*, 3742 (1945).
190. *Dubsky, J. V., A. Langer* and *M. Strnad:* Coll. Czech. Comm. *10*, 111 (1938).
191. *Ducan, L. J. P.*, and *B. F. Clarke:* Ann. Rev. Pharmacol. *5*, 151 (1965).
192. *Duerr, H. H.:* U.S. Patent 2,545,423 (1951); Chem. Abstracts *45*, 6523 (1951).
193. *Dunlop, J. H., R. D. Gillard*, and *G. Wilkinson:* J. Chem. Soc. *1964*, 3160.
194. *Dutta, R. L.*, and *N. R. S. Gupta:* J. Indian Chem. Soc. *38*, 741 (1961).
195. —, and *S. Sarkar:* Sci. Cult. (Calcutta) *30*, 549 (1964).
196. *Duval, D.:* Therapie *14*, 70 (1959).
197. —, Prod. Pharm. *16*, 257 (1961).
198. *Dynamit, A. G.:* German Patent 906,685 (1954); Chem. Abstracts *52*, 11438 (1958).
199. *Eisele, J., W. Federkiel, C. Schuster, R. Gehm*, and *D. Leuchs:* German Patent 1,014,520; Chem. Abstracts *54*, 16852 (1960).
200. *Elslager, E. F.*, and *D. F. Worth:* U.S. Patent 3,074,947 (1963); Chem. Abstracts *59*, 2837 (1963).
201. *Ericks, W. P.:* U.S. Patent 2,350,453 (1944); Chem. Abstracts *38*, 4960 (1944).
202. *Etinger, M. A.:* Sb. Tr. Penzensk. Sel'skokhoz. Inst. *1956*, 301; Chem. Abstracts *54*, 19102 (1960).
203. *Ewan, T.*, and *J. H. Young:* J. Chem. Soc. Ind. *40*, 190 (1921).
204. *Fairweather, H. G. C.:* Brit. Patent 593,019 (1947); Chem. Abstracts *42*, 2995 (1948).
205. *Falcone, A. B., R. L. Moo*, and *E. Shrago:* J. Biol. Chem. *237*, 904 (1962).
206. *Fanshawe, W. J., V. J. Bauer, E. F. Ullman*, and *S. R. Safir:* J. Org. Chem. *29*, 308 (1964).
207. *Farbenindustrie I. G. A.G.:* Belgian Patent 447,750 (1942); Chem. Abstracts *39*, 1000 (1945).
208. *Fernandes, L.*, and *K. Ganapathi:* Proc. Indian Acad. Sci. *28A*, 563 (1948).
209. *Flath, A.:* German Patent 1,058,015 (1959); Chem. Abstracts *59*, 5973 (1961).
210. *Flett, L. H.*, and *N. Y. Scorsdale:* U.S. Patent 2,469,376 (1949); Chem. Abstracts *43*, 6845 (1949).
211. *Flynn, K. G.*, and *D. R. Nenortas:* J. Org. Chem. *28*, 3527 (1963).
212. — — U.S. Patent 3,126,404 (1964); Chem. Abstracts *60*, 16076 (1964).
213. —, and *F. C. Schaefer:* U.S. Patent 3,127,436 (1964); Chem. Abstracts *60*, 15744 (1964); *61*, 2979 (1964).
214. *Fox, A. L.*, and *H. L. Saunders:* U.S. Patent 2,492,855 (1949); Chem. Abstracts *44*, 3298 (1950).
215. *Fraser, D. M.*, and *W. O. Kermack:* Brit. J. Pharmacol. *12*, 16 (1957).
216. *Fraser, G. L.*, and *C. Elmer:* U.S. Patent 2,773,793 (1956); Chem. Abstracts *51*, 7766 (1957).
217. *Fraser, G. P.*, and *W. O. Kermack:* J. Chem. Soc. *1951*, 2682.
218. *Freedman, L., M. Blitz, E. Gunsberg*, and *S. Zak:* J. Lab. Clin. Med. *58*, 662 (1961).
219. —, and *S. L. Shapiro:* Belgian Patent 612,529 (1962); Chem. Abstracts *58*, 8985 (1963).

220. — — Belgian Patent 612,641 (1962); Chem. Abstracts *58,* 8915 (1960).
221. — — U.S. Patent 2,928,768 (1960); Chem. Abstracts *54,* 18565 (1960).
222. *Fukui, T.,* and *Y. Matsuo:* Japan Patent 22,568 (1963); Chem. Abstracts *60,* 4015 (1964).
223. *Fukumara, T., S. Nakayama,* and *R. Namba:* Japan. J. Physiol. *10,* 420 (1960); Chem. Abstracts *55,* 26252 (1961).
224. *Fuller, A. T.:* Biochem. J. *41,* 403 (1947).
225. —, and *H. King:* J. Chem. Soc. *1947,* 963.
226. *Furukawa, M., Y. Seto,* and *S. Toyoshima:* Chem. Pharm. Bull. (Tokyo) *9,* 914 (1961).
227. *Gage, J. C.:* J. Chem. Soc. *1949,* 221.
228. —, and *F. L. Rose:* Ann. Trop. Med. Parasitol. *40,* 333 (1946).
229. *Gaillot, P.,* and *J. Baget:* French Patent 1,045,549 (1953); Chem. Abstracts *52,* 11951 (1958).
230. *Gajewski, F. J.:* U.S. Patent 2,483,969 (1949); Chem. Abstracts *44,* 6159 (1950).
231. *Gale, G. R., A. M. Welch,* and *J. B. Hynes:* J. Pharmacol. Exptl. Therap. *138,* 277 (1962).
232. *Garcia, E. Y.:* J. Philippine Med. Assoc. *26,* 287 (1950).
233. *Geigman, Q. M.:* New Engl. J. Med. *239,* 18; 58 (1948).
234. *Geigy, J. R.,* A.G.: Brit. Patent 546,027 (1942); Chem. Abstracts *37,* 2388 (1943).
235. — Brit. Patent 604,351 (1948); Chem. Abstracts *43,* 4499 (1949).
236. — Swiss Patent 217,129 (1942); Chem. Abstracts *42,* 5683 (1942).
237. — Swiss Patent 223,870 (1942); Chem. Abstracts *43,* 1444 (1949).
238. — Swiss Patent 225,155 (1943); Chem. Abstracts *43,* 2442 (1949).
239. — Swiss Patent 226,853 (1943); Chem. Abstracts *43,* 2786 (1949).
240. — Swiss Patent 232,132 (1944); Chem. Abstracts *43,* 2786 (1949).
241. — Swiss Patent 232,277 (1944); Chem. Abstracts *43,* 4878 (1949).
242. — Swiss Patent 232,278 (1944); Chem. Abstracts *43,* 4878 (1949).
243. — Swiss Patent 232,279 (1944); Chem. Abstracts *43,* 4879 (1949).
244. — Swiss Patent 232,280 (1944); Chem. Abstracts *43,* 4879 (1949).
245. — Swiss Patent 232,281 (1944); Chem. Abstracts *43,* 4879 (1949).
246. — Swiss Patent 232,282 (1944); Chem. Abstracts *43,* 4879 (1949).
247. — Swiss Patent 232,284 (1944); Chem. Abstracts *43,* 4879 (1949).
248. — Swiss Patent 232,285 (1944); Chem. Abstracts *44,* 358 (1950).
249. — Swiss Patent 232,822 (1944); Chem. Abstracts *44,* 358 (1950).
250. — Swiss Patent 233,345 (1944); Chem. Abstracts *44,* 358 (1950).
251. — Swiss Patent 233,346 (1944); Chem. Abstracts *44,* 358 (1950).
252. — Swiss Patent 234,350 (1945); Chem. Abstracts *43,* 4499 (1949).
253. — Swiss Patent 235,189 (1945); Chem. Abstracts *43,* 5199 (1949).
254. — Swiss Patent 235,190 (1945); Chem. Abstracts *43,* 5199 (1949).
255. — Swiss Patent 236,939 (1945); Chem. Abstracts *43,* 7697 (1949).
256. — Swiss Patent 238,832 (1945); Chem. Abstracts *44,* 848 (1950).
257. — Swiss Patent 238,944 (1945); Chem. Abstracts *43,* 4879 (1949).
258. — Swiss Patent 238,945 (1945); Chem. Abstracts *43,* 4880 (1949).
259. — Swiss Patent 238,946 (1945); Chem. Abstracts *43,* 4880 (1949).
260. — Swiss Patent 239,000 (1945); Chem. Abstracts *43,* 7727 (1949).
261. — Swiss Patent 239,001 (1945); Chem. Abstracts *43,* 7632 (1949).
262. — Swiss Patent 240,347 (1946); Chem. Abstracts *44,* 848 (1950).
263. — Swiss Patent 240,348 (1946); Chem. Abstracts *44,* 848 (1950).
264. — Swiss Patent 240,349 (1946); Chem. Abstracts *44,* 848 (1950).

265. — Swiss Patent 240,350 (1946); Chem. Abstracts *44,* 848 (1950).
266. — Swiss Patent 240,353 (1946); Chem. Abstracts *43,* 5216 (1949).
267. — Swiss Patent 240,354 (1946); Chem. Abstracts *43,* 5216 (1949).
268. — Swiss Patent 240,355 (1946); Chem. Abstracts *43,* 5216 (1949).
269. — Swiss Patent 240,356 (1946); Chem. Abstracts *43,* 5216 (1949).
270. — Swiss Patent 240,357 (1946); Chem. Abstracts *43,* 5216 (1949).
271. — Swiss Patent 240,358 (1946); Chem. Abstracts *43,* 5216 (1949).
272. — Swiss Patent 240,359 (1946); Chem. Abstracts *43,* 5216 (1949).
273. — Swiss Patent 240,573 (1946); Chem. Abstracts *43,* 3968 (1949).
274. — Swiss Patent 240,574 (1946); Chem. Abstracts *43,* 3968 (1949).
275. — Swiss Patent 244,766 (1947); Chem. Abstracts *43,* 7727 (1949).
276. — Swiss Patent 244,767 (1947); Chem. Abstracts *43,* 7727 (1949).
277. — Swiss Patent 232,283 (1944); Chem. Abstracts *43,* 4879 (1949).
278. *General Aniline* and *Film Corporation:* Brit. Patent 853,907 (1960); Chem. Abstracts *55,* 15196 (1961).
279. *Georgii, A.:* Intern. Biguanid-Symp., Aachen *1960,* 44; Chem. Abstracts *56,* 12263 (1962).
280. —, u. *H. Mehnert:* Pathol. Anat. Allgem. Pathol. *124,* 278 (1961); Chem. Abstracts *55,* 20206 (1961).
281. *Ghosh, S. P.,* and *A. K. Banerjee:* J. Indian Chem. Soc. *41,* 275 (1964).
282. —, and *A. I. P. Sinha,* J. Indian Chem. Soc. *41,* 330 (1964).
283. *Giles, K. M.,* and *J. E. Harris:* Am. J. Ophthalmol. *48,* 508 (1959).
284. *Gnadt, O. A. F.:* Pharmazie *1,* 103 (1946).
285. *Gontaeva, A. A.:* Med. Parasitol. Parasitic Diseases (USSR) *16,* 64 (1947); Chem. Abstracts *41,* 7537 (1947).
286. *Gortvai, A. F.:* U.S. Patent 2,973,239 (1961); Chem. Abstracts *55,* 17022 (1961).
287. *Goswami, K. N.,* and *S. K. Datta:* Indian J. Phys. *37,* 604 (1963).
288. — — Z. Krist. *120,* 399 (1964).
289. *Goto, Y.,* and *F. D. W. Lukens:* Diabetes *10,* 52 (1961).
290. *Groves, W. W.:* Brit. Patent 511,441 (1939); Chem. Abstracts *34,* 6100 (1940).
291. *Gruen, A.:* U.S. Patent 2,447,175 (1947); Chem. Abstracts *43,* 1814 (1949).
292. — U.S. Patent 2,447,176 (1947); Chem. Abstracts *43,* 1814 (1949).
293. — U.S. Patent 2,447,177 (1947); Chem. Abstracts *43,* 1814 (1949).
294. *Grundland, I.:* Experentia *11,* 118 (1955).
295. *Guha, P. C.,* and *S. S. Guha:* J. Sci. Ind. Research (India) *11B,* 313 (1952).
296. — — J. Sci. Ind. Research (India) *11B,* 317 (1952).
297. — — J. Sci. Ind. Research (India) *11B,* 319 (1952).
298. *Gulland, J. M.,* and *P. E. Macey:* J. Chem. Soc. *1949,* 1257.
299. *Gunther, F. A., G. E. Carmen,* and *M. I. Elliot:* J. Econ. Entomol. *41,* 895 (1948).
300. *Gupta, N. R. S.:* Z. Anorg. Allgem. Chem. *326,* 108 (1963).
301. —, and *P. Ray:* J. Indian Chem. Soc. *37,* 303 (1960).
302. *Gupta, P. K. D., P. Guha,* and *U. P. Basu:* Sci. Cult. (Calcutta) *11,* 704 (1946).
303. *Gupta, P. R.,* and *P. C. Guha:* Current Sci. (India) *17,* 185 (1948).
304. — — Current Sci. (India) *18,* 294 (1949).
305. — — Current Sci. (India) *19,* 312 (1950).
306. —, *B. H. Iyer,* and *P. C. Guha:* Current Sci. (India) *17,* 53 (1948).
307. *Habernickel, V.:* Brit. Patent 787,620 (1957); Chem. Abstracts *52,* 10166 (1958).
308. *Ham, G. P.:* U.S. Patent 2,256,759 (1941); Chem. Abstracts *36,* 49 (1942).
309. *Hansen, J. N.,* and *F. B. Smith:* U.S. Patent 2,304,821 (1942); Chem. Abstracts *37,* 2878 (1943).
310. *Hardy, E. E.:* U.S. Patent 2,627,504 (1953); Chem. Abstracts *47* 5048 (1953).

311. *Harman, M. W.:* U.S. Patent 2,659,755 (1953); Chem. Abstracts *48,* 2313 (1954).
312. *Harris, W. J.:* Australian J. Pharm. *42,* 583 (1961).
313. *Hart, C. A.,* and *C. A. van der Werf:* J. Am. Chem. Soc. *71,* 1875 (1949).
314. *Hechenbleickner, I.:* U.S. Patent 2,768,204 (1956); Chem. Abstracts *51,* 16544 (1957).
315. –, and *D. W. Kaiser:* U.S. Patent 2,768,205 (1956); Chem. Abstracts *51,* 16544 (1957).
316. *Helms, V.,* and *E. D. Weiberg:* Antimicrobial Agents Chemotherapy *1962,* 241.
317. *Hewitt, F.:* Netherlands Patent 79,189 (1955); Chem. Abstracts *51,* 4033 (1957).
318. *Hewitt, R. I., A. Grumble, S. Kushner, S. R. Safir, L. M. Brancone,* and *Y. S. Row:* J. Pharmacol. Exptl. Therap. *96,* 305 (1949).
319. *Heymans, C., J. E. Hyde, P. Terp,* and *G. de Vleeschhouwer:* Arch. Intern. Pharmacodyn. *90,* 140 (1952); Chem. Abstracts *47,* 758 (1953).
320. *Hill, W. H.:* U.S. Patent 2,265,942 (1941); Chem. Abstracts *36,* 2052 (1942).
321. *Hinton, A. J.,* and *V. G. Morgan:* Brit. Patent 851,546 (1960); Chem. Abstracts *55,* 10889 (1961).
322. *Hirt, R. C.,* and *R. G. Schmitt:* Spectrochim. Acta *12,* 127 (1958).
323. *Hodosan, F.,* et *N. Serban:* Bull. Soc. Chim. France *1960,* 133.
324. *Holbrook, A.:* J. Pharm. Pharmacol. *10,* 370 (1958).
325. *Holter, S. N.,* and *W. C. Fernelius:* Inorg. Syn. *7,* 58 (1963).
326. *Hommes, F. A.:* Biochim. Biophys. Acta *77,* 183 (1963).
327. *Horner, L.,* and *K. Kluepfel:* Ann. *591,* 69 (1955).
328. *Houssay, B. A.,* y *J. C. Penhos:* Rev. Soc. Arg. Biol. *34,* 56 (1958); Comp. Rend. Soc. Biol. *152,* 1397 (1950).
329. *Hug, E.:* Ciencia e Invest. (Buenos Aires) *2,* 147 (1946); Chem. Abstracts *42,* 5121 (1948).
330. – Anales Farm. Bioquim. (Buenos Aires), Supl. *17,* 24 (1946); Chem. Abstracts *42,* 5121 (1948).
331. *Hughes, W. B.:* U.S. Patent 2,899,442 (1959); Chem. Abstracts *54,* 1549 (1960).
332. *Inaba, S.:* Japan Patent 14,942 (1963); Chem. Abstracts *60,* 652 (1964).
333. *Inokawa, K.:* Japan Patent 10,320 (1961); Chem. Abstracts *56,* 4626 (1962).
334. *Ishada, N., T. Shiratoi,* and *M. Rikimaru:* J. Antibiotics (Tokyo), Ser. A *15,* 242 (1962).
335. *Jacobs, B. R.,* and *Z. E. Zolles:* Brit. Patent 585,766 (1947); Chem. Abstracts *41,* 6371 (1947).
336. – – Brit. Patent 587,907; Chem. Abstracts *42,* 214 (1948).
337. *Jadzinsky, M. N., C. Pogorelsky, A. de Paula, B. Nusimovich* y *L. B. Fernandez:* Semana Med. (Buenos Aires) *122,* 439 (1963); Chem. Abstracts *59,* 10677 (1963).
338. *Jain, B. C., B. H. Iyer,* and *P. C. Guha:* J. Indian Chem. Soc. *24,* 233 (1947).
339. *James, J. W.,* and *L. F. Wiggins:* Brit. Patent 855,017 (1960); Chem. Abstracts *55,* 13454 (1961).
340. *Jones, S. A.:* Trans. Roy. Soc. Trop. Med. Hyg. *52,* 547 (1958).
341. *Joseph, A. D.,* and *M. N. Jindal:* J. Postgraduate Med. (Bombay) *3,* 225 (1957).
342. *Judith, F. R.,* and *E. D. Weinberg:* J. Bacteriol. *78,* 485 (1959).
343. *Jung, L. M., C. G. Wermuth* et *P. Morand:* Trav. Soc. Pharm. Montpellier *21,* 170 (1961).
344. *Kaiser, D. W.:* U.S. Patent 2,596,930 (1952); Chem. Abstracts *47,* 9562 (1953).
345. –, and *D. Holm-Hansen:* U.S. Patent 2,690,455; Chem. Abstracts *49,* 11701 (1955).
346. *Kalle, A. G.:* German Patent 838,691 (1952); Chem. Abstracts *51,* 11900 (1957).

347. *Karipides, D.*, and *W. C. Fernelius:* Inorg. Syn. *7,* 56 (1963).
348. *Karki, N. T.:* Arch. Intern. Pharmacodyn. *114,* 243 (1958).
349. *Kashiwabara, T. T.:* U.S. Patent 3,199,982 (1965); Chem. Abstracts *63,* 9288 (1965).
350. *Kato, K.:* Nippon Naibumpi Gakkai Zasshi *38,* 587 (1962); Chem. Abstracts *58,* 11844 (1963).
351. *Kaufmann, D.*, and *J. E. O'Hern, jun:* German Patent 1,111,780 (1957); Chem. Abstracts *56,* 3576 (1962).
352. *Kawano, K.*, and *K. Odo:* Nippon Kagaku Zasshi *82,* 1672 (1961); Chem. Abstracts *58,* 11212 (1963).
353. — — Yuki Gosei Kagaku Kyokai Shi *20,* 649 (1962); Chem. Abstracts *59,* 3770 (1963).
354. —, *K. Sugino,* and *K. Odo:* Japan Patent 3874 (1960); Chem. Abstracts *55,* 1455 (1961).
355. *Kazarnovskii, S. N.*, and *N. I. Moschanskaya:* Trudy Gorkovsk. Politekh. Inst. *11,* 62 (1955); Chem. Abstracts *52,* 13539 (1958).
356. — — Zh. Obshch. Khim. *26,* 1948 (1950).
357. *Kendall, S. B.*, and *L. P. Jayner:* Vet. Record *70,* 632 (1958).
358. *Kennerly, G. W.:* U.S. Patent 2,783,210 (1957); Chem. Abstracts *51,* 10572 (1957).
359. *Kimijima, T.*, and *S. Miyama:* J. Soc. Chem. Ind. Japan *46,* suppl. 133 (1943).
360. *King, E. J., I. D. P. Wooton,* and *M. Gilchrist:* Lancet *250,* 886 (1946).
361. *King, H.*, and *I. M. Tonkin:* J. Chem. Soc. *1946,* 1063.
362. *Kitawaki, R.:* Nippon Kagaku Zasshi *85,* 886 (1964); Chem. Abstracts *62,* 16046 (1965).
363. *Krall, L. P.*, and *R. F. Bradley:* Ann. Internal Med. *50,* 586 (1959).
364. —, and *R. Camerini-Davalos:* Proc. Soc. Exptl. Biol. Med. *95,* 345 (1957).
365. *Kretov, A. E.*, and *A. V. Davydov:* Zh. Obshch. Khim. *35,* 1156 (1965).
366. *Kronberg, G.*, u. *K. Stoepel:* Arzneimittel-Forsch. *8,* 470 (1958).
367. *Kruckenburg, W.:* German Patent 1,014,518 (1957); Chem. Abstracts *54,* 9311 (1960).
368. *Kundu, N.:* Sci. Cult. (Calcutta) *15,* 449 (1950).
369. — J. Indian Chem. Soc. *30,* 627 (1953).
370. —, and *P. Ray:* J. Indian Chem. Soc. *29,* 811 (1952).
371. *Kurzer, F.:* Chem. Rev. *56,* 95 (1956).
372. — J. Chem. Soc. *1956,* 1.
373. —, and *K. Douraghi-Zadeh:* Chem. Rev. *67,* 107 (1967).
374. — — J. Chem. Soc. *1965,* 3912.
375. —, and *E. D. Pitchfork:* J. Chem. Soc. *1964,* 3459.
376. — — J. Chem. Soc. *1965,* 6296.
377. — — J. Chem. Soc. (C) *1967,* 1878.
378. — — J. Chem. Soc. (C) *1967,* 1886.
379. —, and *J. R. Powell:* J. Chem. Soc. *1953,* 2537.
380. *Laboratorios Miquel, S. A.:* Spanish Patent 296,600 (1964); Chem. Abstracts *61,* 14588 (1964).
381. *Lazarus, T. S., M. Bradshaw,* and *B. W. Volk:* Diabetes *9,* 118 (1960).
382. *Lecher, H. Z.*, and *F. H. Adams:* U.S. Patent 2,977,329 (1961); Chem. Abstracts *55,* 18159 (1961).
383. —, *W. H. Brunner,* and *F. Pum:* U.S. Patent 3,119,867 (1964); Chem. Abstracts *60,* 14648 (1964).
384. *Leiter, J., J. L. Hartwell, J. S. Kahler, I. Kline,* and *M. J. Shear:* J. Natl. Cancer Inst. *14,* 365 (1953).

385. *Lichtblau, E. I.:* U.S. Patent 2,804,736 (1957); Chem. Abstracts *52,* 4984 (1958).
386. *Linch, A. L.:* U.S. Patent 2,359,864 (1944); Chem. Abstracts *39,* 2078 (1945).
387. *Lipnicki, B.,* and *Z. Wojciechowska:* Arch. Immunol. Terapii Doswiadczalnej *7,* 163 (1959).
388. *Lippmann, H. G.:* Naturwissenschaften *48,* 495 (1962).
389. — Arch. Exptl. Pathol. Pharmakol. *245,* 451 (1963).
390. —, and *J. Lawecki:* Arch. Exptl. Pathol. Pharmakol. *245,* 440 (1963).
391. *Logemann, W., D. Artini* u. *L. Canavesi:* Chem. Ber. *96,* 2914 (1963).
392. — — —, u. *G. Tosolini:* Chem. Ber. *96,* 2909 (1963).
393. *Long, R. S.,* and *S. M. Tsang:* U.S. Patent 2,826,606 (1958); Chem. Abstracts *52,* 10595 (1958).
394. — — U.S. Patent 2,889,317 (1959); Chem. Abstracts *54,* 1870 (1960).
395. — — U.S. Patent 2,898,332 (1959); Chem. Abstracts *54,* 2752 (1960).
396. *Longcope, C.,* and *R. H. Williams:* Proc. Soc. Exptl. Biol. Med. *111,* 775 (1962).
397. *Loo, T. L.:* J. Am. Chem. Soc. *76,* 5096 (1954).
398. *Low, A. M.:* Am. Profess. Pharmacist *12,* 149 (1946).
399. *Lowe, A. J., J. A. Mayse,* and *A. M. Wooler:* Brit. Patent 755,478 (1956); Chem. Abstracts *51,* 5442 (1957).
400. — — — Brit. Patent 798,061 (1958); Chem. Abstracts *53,* 6155 (1959).
401. *Lubowe, I.:* Drug Cosmetic Ind. *81,* 602 (1957); Chem. Abstracts *52,* 5749 (1958).
402. *Lundbaek, K., K. Nielsen,* and *O. J. Rafaelsen:* Ann. N. Y. Acad. Sci. *74,* 419 (1959).
403. *McClellan, P. P.,* and *W. P. Ericks:* U.S. Patent 2,299,245 (1942); Chem. Abstracts *37,* 1877 (1943).
404. — — U.S. Patent 2,369,307 (1945); Chem. Abstracts *39,* 3674 (1945).
405. — — U.S. Patent 2,369,309 (1945); Chem. Abstracts *40,* 2998 (1946).
406. *McDonald, H. J.,* and *J. E. Dalidowicz:* Biochemistry *1,* 1187 (1962).
407. *McDowall, M. A.,* and *H. Waal:* Brit. J. Pharmacol. *14,* 527 (1959).
408. *McGeachin, R. L.:* J. Am. Chem. Soc. *75,* 973 (1953).
409. *McGregor, J. M.:* U.S. Patent 2,448,448 (1948); Chem. Abstracts *43,* 868 (1949).
410. — U.S. Patent 2,458,397 (1949); Chem. Abstracts *43,* 2442 (1949).
411. *McKay, A. F.:* Brit. Patent 987,252; Chem. Abstracts *63,* 1800 (1965).
412. *McKee, R. H.:* Am. Chem. J. *36,* 208 (1906).
413. *McLaughlan, J. M., K. G. Shenoy,* and *J. A. Campbell:* J. Pharm. Sci. *50,* 59 (1961).
414. *McLeod, J. A.,* and *J. W. McLeod:* J. Pathol. Bacteriol. *77,* 219 (1959).
415. *Maegraith, B. G., A. R. D. Adams, J. D. King, R. H. Townsend, T. Davey,* and *R. E. Harvard:* Ann. Trop. Med. Parasitol. *39,* 232 (1945).
416. *Mamalis, P., J. Green,* and *D. McHale:* J. Chem. Soc. *1960,* 229.
417. — —, *D. J. Outred,* and *M. Rix:* J. Chem. Soc. *1962,* 3915.
418. —, *D. McHale,* and *K. J. Stevens:* Brit. Patent 814,563 (1959); Chem. Abstracts *53,* 21674 (1959).
419. *Manuelli, C.:* Ann. Chim. Applic. *23,* 231 (1933).
420. *Marckwald, W.:* Ann. *286,* 361 (1895).
421. *Marshall, E. K. jun.,* and *E. H. Dearborn:* J. Pharmacol. Exptl. Therap. *88,* 187 (1946).
422. *Marshall, P. B.:* Nature *160,* 463 (1947).
423. *Marson, H. W.,* and *J. P. English:* U.S. Patent 2,675,375 (1954); Chem. Abstracts *49,* 6322 (1955).

424. *Marxer, A.*: Swiss Patent 383,949 (1965); Chem. Abstracts *62,* 16136 (1965).
425. *Matsuda, K.*: Naika Hokan *5,* 667 (1958); Chem. Abstracts *53,* 10524 (1959).
426. *May, E. L.*: J. Org. Chem. *12,* 437 (1947).
427. — J. Org. Chem. *12,* 443 (1947).
428. —, and *E. Mosettig*: J. Org. Chem. *12,* 869 (1947).
429. *Mehnert, H.*: Deut. Med. Wochschr. *83,* 1273 (1958).
430. — Chemotherapia *2,* 262 (1961).
431. *Mehta, D. J., C. V. Deliwala, M. H. Shah,* and *V. K. Sheth*: Arch. Intern. Pharmacodyn. *138,* 480 (1962).
432. *Melander, B.*: Toxicol. Appl. Pharmacol. *2,* 474 (1960).
433. *Meuly, W. C.*: U.S. Patent 2,910,497 (1959); Chem. Abstracts *54,* 3210 (1960).
434. *Michel, W.*: Intern. Biguanid-Symp., Aachen *1960,* 111; Chem. Abstracts *56,* 12279 (1962).
435. *Michelsen, K.*: Acta Chem. Scand. *19,* 1175 (1965).
436. *Milks, J. E., G. W. Kennerly,* and *J. H. Polevy*: J. Am. Chem. Soc. *84,* 2529 (1962).
437. *Miller, J.*: J. Chem. Soc. *1949,* 2722.
438. *Miller, W. A.*: Poultry Sci. *36,* 579 (1957); Chem. Abstracts *51,* 15825 (1957).
439. *Mingoia, Q.,* e *P. C. Ferreira*: Anais Faculdade Farm. Odontal. Univ. Sao Paulo *7,* 43 (1949); Chem. Abstracts *45,* 1972 (1951).
440. *Misba, N. K.,* u. *G. B. Mitra*: Z. Krist. *119,* 155 (1963).
441. *Modest, E. J.*: J. Org. Chem. *21,* 1 (1956).
442. — In: Heterocyclic Compounds, Vol. 7, p. 697 seq. *R. C. Elderfield* Ed. New York: John Wiley and Son, Inc. 1961.
443. —, and *P. Levine*: J. Org. Chem. *21,* 14 (1956).
444. *Monsanto Chemicals Ltd.*: French Patent 1,394,711; Chem. Abstracts *63,* 17972 (1965).
445. — Netherlands pat. applic. 289,283 (1965); Chem. Abstracts *63,* 9867 (1965).
446. *Moore, L. P.,* and *W. P. Ericks*: U.S. Patent 2,323,409 (1943).
447. — — U.S. Patent 2,453,333 (1948); Chem. Abstracts *43,* 1799 (1949).
448. *Morgan, W. L.,* and *E. D. McLeod*: U.S. Patent 2,362,768 (1944); Chem. Abstracts *39,* 5371 (1945).
449. *Mshvidobadze, A. E., B. I. Chumburidze,* and *O. V. Sardzhveladze*: Aptechn. Delo *12,* 36 (1963); Chem. Abstracts *61,* 11853 (1964).
450. *Mueller, D., H. Guethert* u. *H. Benzold*: Zentr. Allgem. Pathol. Pathol. Anat. *104,* 124 (1963); Chem. Abstracts *59,* 13829 (1963).
451. *Mueller, F. W. H.*: U.S. Patent 2,584,362 (1952); Chem. Abstracts *46,* 3436 (1952).
452. — U.S. Patent 2,591,590 (1952); Chem. Abstracts *46,* 6024 (1952).
453. *Mushkin, Y. I.,* and *A. I. Finkelshtein*: Zh. Organ. Khim. *1,* 721 (1965); Chem. Abstracts *63,* 8196 (1965).
454. *Nagy, D. E.*: U.S. Patent 2,309,661 (1943).
455. — U.S. Patent 2,446,980 (1948).
456. — U.S. Patent 2,455,896 (1948); Chem. Abstracts *43,* 2229 (1949).
457. — U.S. Patent 2,455,897 (1948); Chem. Abstracts *43,* 1799 (1949).
458. — U.S. Patent 2,621,102 (1952); Chem. Abstracts *47,* 5709 (1953).
459. *Nakamura, T.,* and *Y. Aida*: Jap. Patent 4891 (1963); Chem. Abstracts *59,* 11333 (1963).
460. *Nakano, T.*: Kurume Med. *5,* 101 (1958); Chem. Abstracts *53,* 8436 (1959).
461. *Nandi, K. N.,* and *M. A. Phillips*: Chem. Ind. *1958,* 719.
462. *Nandy, S. C.,* and *S. N. Sen*: Bull. Central Leather Res. Inst. Madras (India) *6,* 315 (1960); Chem. Abstracts *54,* 12625 (1960).

463. *Natarajan, S.*, and *H. L. Bami:* Current Sci. (India) *16*, 156 (1947).
464. *Neelakantan, L.:* J. Org. Chem. *22*, 1584 (1957).
465. — J. Org. Chem. *22*, 1587 (1957).
466. —, *B. H. Iyer*, and *P. C. Guha:* J. Indian Chem. Soc. *29*, 131 (1952).
467. *Neidlein, R., G. Kluegel* u. *U. Lebert:* Pharm. Ztg., No. 20, 657 (1965); Chem. Abstracts *63*, 6788 (1965).
468. *Neuman, R. E.*, and *A. A. Tytell:* Proc. Soc. Exptl. Biol. Med. *110*, 622 (1962).
469. — — Proc. Soc. Exptl. Biol. Med. *110*, 627 (1962).
470. — — Proc. Soc. Exptl. Biol. Med. *112*, 57 (1963).
471. *Newman, H.*, and *E. L. Moon:* J. Med. Chem. *8*, 702 (1965).
472. *Nielsen, R. L., H. E. Swanson, D. C. Tanner, R. M. Williams*, and *M. O'Connell:* A. M. A. Arch. Internal Med. *101*, 211 (1958).
473. *Nyberg, D. D.*, and *B. E. Christensen:* J. Am. Chem. Soc. *78*, 781 (1956).
474. *Oji, N.:* Nippon Naibumpi Gakkai Zasshi *35*, 1004 (1959); Chem. Abstracts *54*, 19950 (1960).
475. *Oldham, W.:* U.S. Patent 2,309,663 (1943).
476. — U.S. Patent 2,320,882 (1943).
477. — U.S. Patent 2,344,784 (1944).
478. *Olipin, H. C., P. B. Law*, and *S. A. Gibson:* U.S. Patent 2,474,909 (1949); Chem. Abstracts *43*, 7713 (1949).
479. *Opitz, K.:* Klin. Wochschr. *40*, 56 (1962).
480. —, u. *A. Loeser:* Deut. Med. Wochsch. *87*, 105 (1962).
481. *Osterloh, G.:* Internat. Biguanid-Symp., Aachen *1960*, 8; Chem. Abstracts *56*, 13504 (1962).
482. *Ostrogovich, A.:* Bull. Soc. Stiinte Bucharesti *19*, 641 (1910); Chem. Zentral. *1910 II*, 1890.
483. *Otto, H.:* Internat. Biguanid-Symp., Aachen *1960*, 97; Chem. Abstracts *56*, 12279 (1962).
484. *Overberger, C. G., F. W. Michelotti*, and *P. M. Carabateas:* J. Am. Chem. Soc. *79*, 941 (1957).
485. —, and *S. L. Shapiro:* J. Am. Chem. Soc. *76*, 93 (1954).
486. — — J. Am. Chem. Soc. *76*, 1061 (1954).
487. *Owens, T. P.*, u. *L. S. Malowan:* Arzneimittel-Forsch. *9*, 655 (1959).
488. *Oxley, P.*, and *W. F. Short:* J. Chem. Soc. *1951*, 1252.
489. *Paden, J. H., K. C. Martin*, and *R. C. Swain:* Ind. Eng. Chem. *39*, 952 (1947).
490. *Paintal, A. S.:* J. Physiol. *126*, 271 (1954).
491. — Quart. J. Exptl. Physiol. *42*, 56 (1957).
492. *Palmas, S.*, i *R. Rimini:* Studi Sassaresi, Sezione *1*, *39*, 239 (1961).
493. *Parke Davis* and *Co.:* Brit. Patent 986,811 (1965); Chem. Abstracts *63*, 13294 (1965).
494. *Paul, S. P.*, and *A. N. Bose:* Indian J. Physiol. Pharmacol. *6*, 3846 (1962).
495. — — Indian J. Physiol. Pharmacol. *7*, 55 (1963).
496. — —, and *U. P. Basu:* Indian J. Chem. *1*, 218 (1963).
497. *Pellizzari, G.:* Atti accad. Lincei *30I*, 224 (1921).
498. — Gazz. Chim. Ital. *53*, 384 (1923).
499. *Peng, M. T.*, and *S. C. Wang:* Arch. Intern. Pharmacodyn. *140*, 695 (1962).
500. *Penhos, J. C.*, y *J. A. Blaquier:* Rev. Soc. Arg. Biol. *34*, 21 (1958); Compt. Rend. Soc. Biol. *152*, 1392 (1950).
501. *Persch, W.*, and *H. G. Greve:* German Patent 1,024,516 (1958); Chem. Abstracts *54*, 19730 (1960).
502. — — German Patent 1,080,111 (1960); Chem. Abstracts *56*, 485 (1962).

503. *Pessel, L.:* U.S. Patent 2,690,408 (1954); Chem. Abstracts *49,* 1530 (1955).
504. *Petersen, S., H. F. Piepenbrink,* and *O. Bayer:* German Patent 968,041 (1958). Chem. Abstracts *53,* 16602 (1959).
505. *Pfeiffer, E. F.,* u. *H. G. Rossenbeck:* Klin. Wochschr. *39,* 71 (1961).
506. *Piccinini, F., E. Marazzi Uberti* i *I. Lucattelli:* Farmaco (Pavia), Ed. Sci. *15,* 521 (1960).
507. *Pichappa, C. V.,* and *E. R. B. Shanmugasundarum:* J. Bacteriol. *81,* 831 (1961).
508. *Pignard, A.:* Ann. Biol. Chem. *20,* 325 (1962).
509. *Pilcher, K., K. F. Soike,* and *F. Trosper:* Antibiot. Chemotherapy *11,* 381 (1961).
510. *Poddar, S. N.:* Sci. Cult. (Calcutta) *29,* 50 (1963).
511. *Pomeranze, J., H. Fujiy,* and *G. T. Mouratoff:* Proc. Soc. Exptl. Biol. Med. *95,* 193 (1957).
512. *Price, S. A., P. Mamalis, D. McHale,* and *J. Green:* Brit. J. Pharmacol. *15,* 243 (1960).
513. *Proske, G., H. Mueckter, G. Osterloh,* and *H. W. Schrader-Beielstein:* German Patent 1,093,349 (1960); Chem. Abstracts *55,* 19801 (1961).
514. —, *G. Osterloh, R. Beckmann, F. Lagler, G. Michael* u. *H. Mueckter:* Arzneimittel-Forsch. *12,* 314 (1962).
515. *Puetzer, B.:* U.S. Patent 2,191,860 (1940); Chem. Abstracts *34,* 4528 (1940).
516. — U.S. Patent 2,213,474 (1940); Chem. Abstracts *35,* 1067 (1941).
517. *Pushkareva, Z. V., V. I. Shishkina,* and *L. V. Varynkhina:* Doklady Akad. Nauk. SSSR *92,* 89 (1953); Chem. Abstracts *44,* 10653 (1950).
518. *Rackmann, K.:* Ann. *376,* 163 (1910).
519. — Ann. *376,* 170 (1910).
520. *Rafaelsen, O. J.:* Metab., Clin. Exptl. *8,* 195 (1959).
521. *Raman, K., M. Raghavan,* and *P. C. Guha:* J. Indian Inst. Sci. *35A,* 247 (1953).
522. *Raman, T. S.,* and *E. R. B. Shanmugasundarum:* Ann. Biochem. Exptl. Med. (Calcutta) *22,* 21 (1962).
523. *Rangam, C. M.,* and *R. R. Bhagwat:* Indian J. Med. Research *48,* 549 (1960).
524. *Rathke, B.:* Ber. *12,* 776 (1879).
525. *Ray, A. K.:* Z. Anal. Chem. *156,* 18 (1957).
526. *Ray, A. P., C. P. Nair, M. K. Menon,* and *B. G. Misra:* Indian J. Malariol. *8,* 209 (1954).
527. *Ray, M. M.:* Sci. Cult. (Calcutta) *30,* 190 (1964).
528. — J. Inorg. Nucl. Chem. *27,* 2193 (1965).
529. —, *K. De,* and *S. N. Poddar:* Indian J. Chem. *3,* 228 (1965).
530. *Ray, P.:* Nature *152,* 694 (1943).
531. — J. Indian Chem. Soc. *32,* 141 (1955).
532. — Chem. Rev. *61,* 313 (1961).
533. — Inorg. Syn. *7,* 6 (1963).
534. *Rembarz, G., H. Brandner* u. *H. Finger:* J. Prakt. Chem. *26,* 314 (1964).
535. *Richter, G. Vegyeszeti Gyar:* Brit. Patent 676,024 (1952).
536. *Ridi, M.,* i *S. Checchi:* Ann. Chim. (Rome) *43,* 807 (1953).
537. — — Ann. Chim. (Rome) *44,* 28 (1954).
538. *Robbins, M. L., A. R. Bourke,* and *P. K. Smith:* J. Immunol. *64,* 431 (1950).
539. *Robertson, A. E.,* and *N. J. Roselle:* U.S. Patent 2,469,745 (1945); Chem. Abstracts *43,* 6231 (1949).
540. *Robertson, G. I.:* Trans. Roy. Soc. Trop. Med. Hyg. *51,* 457 (1957).
541. — Trans. Roy. Soc. Trop. Med. Hyg. *51,* 488 (1957).
542. *Robertson, J. E.:* French Patent 1,320,255 (1963); Chem. Abstracts *59,* 10010 (1963).

543. *Robinson, J. G.:* Trans. Inst. Rubber Ind. *17,* 33 (1941); Rubber Chem. Technol. *15,* 128 (1942).
544. *Rochlin, P., D. B. Murphy,* and *S. Helf:* J. Am. Chem. Soc. *76,* 1451 (1954).
545. *Rose, F. L.:* Endeavour *5,* 65 (1946).
546. — J. Chem. Soc. *1951,* 2770 (For Summary).
547. — J. Chem. Soc. *1952,* 3448.
548. — Brit. Patent 550,538 (1943); Chem. Abstracts *38,* 1610 (1944).
549. —, and *G. Swain:* J. Chem. Soc. *1956,* 4422.
550. — — Brit. Patent 705,838 (1954); Chem. Abstracts *50,* 1082 (1956).
551. *Rose, W.,* u. *C. Jaehn:* Aerztl. Forsch. *15,* 493 (1961); Chem. Abstracts *56,* 9355 (1962).
552. *Rosenthaler, L.:* Biochem. Z. *207,* 298 (1929).
553. *Roskova, H.:* Compt. Rend. Soc. Biol. *142,* 172 (1948).
554. *Rosowsky, A., H. K. Protpapa, P. J. Burke,* and *E. J. Modest:* J. Org. Chem. *29,* 2881 (1964).
555. *Roth, B., R. B. Burrows,* and *G. H. Hitchings:* J. Med. Chem. *6,* 370 (1963).
556. *Roy, A. C.,* and *P. C. Guha:* J. Sci. Ind. Res. (India) *9B,* 242 (1950).
557. —, *S. S. Guha, N. K. Keshavamurthy, G. R. Chandrasekhar, K. P. Menon,* and *P. C. Guha:* J. Sci. Ind. Res. (India) *12B,* 474 (1953).
558. *Royer, R.:* J. Chem. Soc. *1949,* 1665.
559. *Rumpf, P. H.:* French Patent 948,328 (1949); Chem. Abstracts *45,* 5723 (1951).
560. *Ryan, W. H.,* and *V. K. Walworth:* U.S. Patent 2,868,077 (1959); Chem. Abstracts *53,* 7842 (1959).
561. *Ryley, J. F.:* Brit. J. Pharmacol. *8,* 424 (1953).
562. *Safir, S. A., S. Kushner, L. M. Brancone,* and *Y. Subba Row:* J. Org. Chem. *13,* 924 (1948).
563. *Sammul, O. R., W. L. Brannon,* and *A. L. Hayden:* J. Assoc. Offic. Agr. Chemists *47,* 918 (1964); Chem. Abstracts *61,* 14467 (1964).
564. *Sandoz, Ltd.:* Brit. Patent 611,235 (1948); Chem. Abstracts *43,* 3206 (1949).
565. — Swiss Patent 261,048 (1949); Chem. Abstracts *44,* 3719 (1950).
566. — Swiss Patent 261,049 (1949); Chem. Abstracts *44,* 3719 (1950).
567. — Swiss Patent 261,050 (1949); Chem. Abstracts *44,* 3719 (1950).
568. — Swiss Patent 261,051 (1949); Chem. Abstracts *44,* 3719 (1950).
569. *Sarma, B. D.:* J. Indian Chem. Soc. *29,* 217 (1952).
570. *Sarma, P. S.:* Indian Patent 57,268 (1958); Chem. Abstracts *52,* 20770 (1958).
571. — Res. Ind. *8,* 221 (1963); Chem. Abstracts *60,* 15049 (1964).
572. *Sato, M.:* J. Pharm. Soc. Japan *69,* 303 (1949); Chem. Abstracts *44,* 3931 (1950).
573. *Schaaf, A. van der, F. H. J. Jaartsueld,* and *P. H. A. M. van Maanen:* Tijdschr. Diergeneesk. *85,* 185 (1960); Chem. Abstracts *57,* 17197 (1962).
574. *Schalit, S.,* and *R. A. Cutler:* J. Org. Chem. *24,* 573 (1959).
575. *Schmidt, L. H., H. B. Hughes,* and *C. C. Smith:* J. Pharmacol. Exptl. Therap. *90,* 233 (1947).
576. *Schneider, F.:* Z. Physiol. Chem. *334,* 26 (1963).
577. *Schroeder, H. A., H. M. Perry Jr., E. M. Menhard,* and *E. G. Denis:* J. Lab. Clin. Med. *40,* 416 (1955).
578. *Schuette, E.:* Z. Physiol. Chem. *279,* 52 (1943).
579. *Schwalenstocker, H. J.:* U.S. Patent 2,668,113 (1954); Chem. Abstracts *48,* 5700 (1954).
580. *Scott, F. L.:* J. Org. Chem. *22,* 156 (1957).
581. *Sen, A. B.,* and *S. K. Gupta:* J. Indian Chem. Soc. *40,* 578 (1963).
582. —, and *P. R. Singh:* J. Indian Chem. Soc. *39,* 41 (1962).

583. *Sen, A. K., N. K. Ray,* and *U. P. Basu:* J. Sci. Ind. Res. (India) *11B,* 322 (1952).
584. — — — J. Sci. Ind. Res. (India) *11B,* 324 (1952).
585. —, *A. Raychandhuri,* and *U. P. Basu:* J. Sci. Ind. Res. (Ind a) *11B,* 325 (1952).
586. *Sen, D.:* Sci. Cult. (Calcutta) *28,* 292 (1962).
587. *Senior, N.:* Brit. Patent 843,676 (1960); Chem. Abstracts *55,* 4429 (1961).
588. —, and *A. Sweeny:* Brit. Patent 815,800 (1959); Chem. Abstracts *54,* 1421 (1960).
589. *Serafin, B.,* and *T. Urbanski:* Roczniki Chem. *36,* 679 (1962).
590. — —, and *J. Zylowski:* Nitro-Compounds, Proc. Intern. Symp., Warsaw, 1963, 469; Chem. Abstracts *63,* 16346 (1965).
591. *Sewel, P.,* and *F. Hawking:* Brit. J. Pharmacol. *5,* 239 (1950).
592. *Shaknazarova, N. G.:* Khim. i. Med. *1956,* 117; Chem. Abstracts *52,* 1487 (1958).
593. *Shapiro, S. L.,* and *L. Freedman:* U.S. Patent 2,937,170 (1960); Chem. Abstracts *54,* 18565 (1960).
594. — — U.S. Patent 2,961,377 (1960); Chem. Abstracts *55,* 12784 (1961).
595. — — U.S. Patent 3,057,780 (1962); Chem. Abstracts *58,* 418 (1963).
596. — — U.S. Patent 3,098,008 (1963); Chem. Abstracts *60,* 2793 (1964).
597. — — U.S. Patent 3,131,128 (1964); Chem. Abstracts *61,* 4324 (1964).
598. — — U.S. Patent 3,147,271 (1964); Chem. Abstracts *62,* 1634 (1965).
599. —, *E. S. Isaacs, V. A. Parrino,* and *L. Freedman:* J. Org. Chem. *26,* 68 (1961).
600. —, and *C. G. Overberger: J.* Am. Chem. Soc. *76,* 97 (1954).
601. —, *V. A. Parrino,* and *L. Freedman:* J. Am. Pharm. Assoc., Sci. Ed. *46,* 689 (1957).
602. — — — J. Am. Chem. Soc. *81,* 2220 (1959).
603. — — — J. Am. Chem. Soc. *81,* 3728 (1959).
604. — — — J. Am. Chem. Soc. *81,* 4635 (1959).
605. — — — J. Org. Chem. *25,* 379 (1960).
606. — — — J. Org. Chem. *25,* 384 (1960).
607. — —, *K. Geiger, S. Korobin,* and *L. Freedman:* J. Am. Chem. Soc. *79,* 5064 (1957).
608. — —, *E. Rogow,* and *L. Freedman:* J. Am. Chem. Soc. *81,* 3725 (1959).
609. *Shaw, J. T.,* and *F. J. Cross:* J. Org. Chem. *24,* 1809 (1959).
610. *Shepherd, H. G. jun.,* and *H. J. McDonald:* Clin. Chem. *4,* 496 (1958).
611. — — Proc. Soc. Exptl. Biol. Med. *102,* 390 (1959).
612. *Shirai, K.,* and *K. Sugino:* J. Org. Chem. *25,* 1045 (1960).
613. *Short, W. F., G. I. Hobday,* and *P. Oxley:* Brit. Patent 610,379 (1948); Chem. Abstracts *43,* 4292 (1949).
614. — — — U.S. Patent 2,473,112 (1949); Chem. Abstracts *43,* 7040 (1949).
615. *Shultz, R. C.:* J. Am. Pharm. Assoc. *38,* 84 (1949).
616. *Siedlanowska, H.:* Acta Polon. Pharm. *21,* 33 (1964); Chem. Abstracts *62,* 14423 (1965).
617. — Farm. Polska *19,* 214 (1963); Chem. Abstracts *60,* 1538 (1964).
618. *Siest, G., F. Roos* et *J. J. Gabou:* Bull. Soc. Pharm. Nancy *58,* 29 (1963); Chem. Abstracts *60,* 2722 (1964).
619. *Simons, J. K.:* U.S. Patent 2,273,382 (1942); Chem. Abstracts *36,* 3875 (1942).
620. *Singh, J., A. P. Ray,* and *C. P. Nair:* Indian J. Malariol. *3,* 387 (1949).
621. — — — Indian J. Malariol. *10,* 85 (1956).
622. *Sirsi, M., R. R. Rao,* and *N. N. De:* Current Sci. (India) *19,* 317 (1950).
623. *Skillman, T. G., F. A. Kruger,* and *G. J. Hamivi:* Diabetes *8,* 274 (1959).
624. *Slifkin, S. C.:* U.S. Patent 2,542,716 (1951); Chem. Abstracts *45,* 6521 (1951).

625. *Slopek, S., H. Mordarska, M. Mordarski, T. Urbanski, B. Skowronska-Serafin,* and *H. Dabrowska:* Bull. Acad. Polon. Sci. *6,* 355 (1958); Chem. Abstracts *52,* 20604 (1958).
626. *Slotta, K. H.,* and *R. Tschesche:* Ber. *62,* 1398 (1929).
627. *Smith, C. C., J. Ihrig,* and *R. Menne:* Am. J. Trop. Med. Hyg. *10,* 694 (1961).
628. *Smith, F. B.,* and *J. N. Hansen:* U.S. Patent 2,367,534 (1945); Chem. Abstracts *39,* 3112 (1945).
629. *Smith, G. B. L., V. J. Sabetta,* and *O. F. Steinbach jun.:* Ind. Eng. Chem. *23,* 1124 (1931).
630. *Smolka, A.,* and *A. Friedreich:* Monatsh. *9,* 228 (1868).
631. — — Monatsh. *10,* 87 (1869).
632. *Société des Usines Chimiques Rhone-Poulenc:* Brit. Patent 622,266 (1949); Chem. Abstracts *43,* 7507 (1949).
633. — Brit. Patent 643,000 (1950).
634. — French Patent 948,375 (1949); Chem. Abstracts *45,* 5723 (1951).
635. *Société d'Etudes Scientifiques et Industrielles de l'Ile-de-France:* French Patent M 1429 (1962); Chem. Abstracts *58,* 3279 (1963).
636. *Société pour l'Industrie Chimique à Basle:* Brit. Patent 587,263 (1947). Chem. Abstracts *41,* 7642 (1947).
637. — Brit. Patent 616,694 (1949); Chem. Abstracts *43,* 5978 (1949).
638. — Swiss Patent 205,525 (1939); Chem. Abstracts *35,* 2534 (1941).
639. *Soeling, H. D., F. Andreu-Kem, H. Werchau* u. *W. Creutzfeldt:* Klin. Wochschr. *39,* 1080 (1962).
640. —, u. *W. Creutzfeldt:* Inter. Biguanid Symp., Aachen *1960,* 17; Chem. Abstracts *56,* 12262 (1962).
641. —, *H. Werchau* u. *W. Creutzfeldt:* Arch. Exptl. Pathol. Pharmacol. *244,* 290 (1963).
642. *Sokolovskaya, S. V., V. N. Sokolova,* and *O. Y. Magidson:* J. Gen. Chem. USSR (English Transl.) *27,* 839 (1957).
643. — — — J. Gen. Chem. USSR (English Transl.) *27,* 1103 (1957).
644. — — — J. Gen. Chem. USSR (English Transl.) *27,* 2030 (1957).
645. *Solomon, J., M. Madgearu,* and *M. Prodescu:* Farmacia (Bucharest) *9,* 167 (1961); Chem. Abstracts *56,* 14400 (1962).
646. *Solonskaya, N. T.,* and *L. S. Sokol:* Farmatsevt. Zh. (Kiev) *15,* 13 (1960); Chem. Abstracts *55,* 5384 (1961).
647. *Somogyi, L., Z. Gyorgydeak,* and *R. Bognar:* Tetrahedron Letters *1966,* 871.
648. *Spacu, P.,* and *I. Albescu:* Acad. Rep. Populare Romine, Studii Cercetari Chim. *8,* 85 (1960); Chem. Abstracts *54,* 20656 (1960).
649. —, *M. Brezeanu, C. Gheorgiu, O. Constantinescu,* and *I. Pascaru:* Rev. Roumaine Chim. *9,* 801 (1964); Chem. Abstracts *63,* 12543 (1965).
650. —, *C. Gheorgiu,* and *E. Cristurean:* Rev. Roumaine Chim. *9,* 265 (1964); Chem. Abstracts *61,* 9159 (1964).
651. *Spinks, A.:* Ann. Trop. Med. Parasitol. *41,* 30 (1947).
652. —, and *M. M. Tottey:* Ann. Trop. Med. Parasitol. *39,* 220 (1945).
653. — — Ann. Trop. Med. Parasitol. *40,* 101 (1946).
654. — — Ann. Trop. Med. Parasitol. *40,* 153 (1946).
655. *Sprung, J. A.:* U.S. Patent 2,704,710 (1955); Chem. Abstracts *49,* 8019 (1955).
656. *Standard Oil Development Co.:* Brit. Patent 619,018 (1949); Chem. Abstracts *43,* 5799 (1949).
657. *Steinbach, M.:* Bibliotheca Haematol., Suppl. Acta Haematol. *1956,* 323.
658. —, et *M. Quentin:* Arch. Intern. Pharmacodyn. *110,* 10 (1957).
659. *Steiner, D. F.,* and *R. H. Williams:* Biochem. et Biophys. Acta *30,* 329 (1958).

660. *Sterne, J.:* Therapie *13,* 650 (1958).
661. — Congr. Fédération Intern. Diabète *4,* Geneva *1961,* 1, 712; Chem. Abstracts *57,* 13153 (1962).
662. — Wien. Med. Wochschr. *113,* 599 (1963); Chem. Abstracts *60,* 2233 (1964).
663. *Stewart, W. K.,* and *L. W. Constable:* Lancet *1961,* 523.
664. *Stokes, J. A.:* U.S. Patent 2,577,593 (1951); Chem. Abstracts *46,* 2309 (1952).
665. *Stowers, J. N.:* Proc. Roy. Soc. Med. *53,* 603 (1960).
666. *Streuli, C. A.,* and *J. A. McCormack:* Unpublished results quoted in J. J. *Cincotta* and *R. Feinland,* Anal. Chem. *34,* 774 (1962) (i. e. ref. 110).
667. *Stukov, I. T., N. A. Sychra,* and *B. P. Sirmov:* J. Chem. Ind. USSR *18,* No. 17, 22 (1941); Chem. Abstracts *38,* 5410 (1944).
668. *Sugino, K.:* J. Chem. Soc. Japan *60,* 351 (1939).
669. — J. Chem. Soc. Japan *60,* 411 (1939).
670. —, *Y. Aiya,* and *K. Ariga:* J. Soc. Chem. Ind. Japan *46,* 573 (1943).
671. —, and *S. Idzumi:* J. Chem. Soc. Japan *65,* 265 (1944).
672. —, and *M. Ogawa:* J. Electrochem. Assoc. Japan *60,* 292 (1938); Rev. Phys. Chem. Japan *13,* 58 (1939).
673. —, and *M. Yamashita:* J. Chem. Soc. Japan *65,* 271 (1944).
674. *Supniewski, J.,* and *T. Chrusciel:* Arch. Immunol. Terapii Doswiadczalnej *2,* 1 (1954); Chem. Abstracts *52,* 4031 (1958).
675. — — Bull. Acad. Polon. Sci. *2,* 29 (1954); Chem. Abstracts *48,* 13058 (1954).
676. —, and *J. Krupinska:* Bull. Acad. Polon. Sci. *2,* 161 (1954); Chem. Abstracts *49,* 14186 (1955).
677. *Sutton, B. M.:* U.S. Patent 2,934,535 (1960); Chem. Abstracts *54,* 24827 (1960).
678. *Suzuki, A., M. Inone, S. Suganuma,* and *T. Ohara:* Kobe Ika Daigaku Kiyo *13,* 658 (1958); Chem. Abstracts *56,* 4055 (1962).
679. *Swain, R. C.:* U.S. Patent 2,251,234 (1941); Chem. Abstracts *35,* 6846 (1941).
680. *Swales, W. E.:* Can. J. Comp. Med. Vet. Sci. *11,* 123 (1947).
681. *Swaminathan, S.:* Indian Patent 37,045 (1948).
682. *Szabo, L., L. Szporny,* and *O. Clauder:* Acta Pharm. Hung. *31,* 163 (1961).
683. *Szulczewski, D. A., C. M. Shearer,* and *A. J. Aguiar:* J. Pharm. Sci. *53,* 1156 (1964).
684. *Takagi, K.,* and *T. Ueda:* Chem. Pharm. Bull. Japan *12,* 607 (1964).
685. *Takase, S.,* and *H. Matsui:* Sci. Repts. Osaka University *13,* 21 (1964); Chem. Abstracts *63,* 13140 (1965).
686. *Takimoto, M.:* Kogyo Kagaku Zasshi *64,* 1456 (1960); Chem. Abstracts *57,* 1456 (1962).
687. — Nippon Kagaku Zasshi *85,* 159 (1964); Chem. Abstracts *61,* 2937 (1964).
688. — Nippon Kagaku Zasshi *85,* 168 (1964); Chem. Abstracts *61,* 2937 (1964).
689. —, and *T. Yatsuo:* Kogyo Kagaku Zasshi *63,* 1938 (1960).
690. — — Kogyo Kagaku Zasshi *63,* 1941 (1960); Chem. Abstracts *57,* 1555 (1962).
691. *Tanner, E.:* German Patent 1,059,602 (1959); Chem. Abstracts *55,* 15966 (1961).
692. *Taube, K.,* and *K. Bloeckmann:* German Patent 954,686 (1956); Chem. Abstracts *53,* 1760 (1959).
693. *Tendick, F. H.,* and *J. H. Burckhalter:* J. Am. Chem. Soc. *72,* 1862 (1950).
694. *Thiele, J.:* Ann. *273,* 133 (1893).
695. *Thurston, J. P.:* Brit. J. Pharmacol. *5,* 409 (1950).
696. *Thurston, J. T.:* U.S. Patent 2,305,118 (1943); Chem. Abstracts *37,* 2747 (1943).
697. — U.S. Patent 2,309,679 (1943); Chem. Abstracts *37,* 3769 (1943).
698. — U.S. Patent 2,390,476 (1945); Chem. Abstracts *40,* 1544 (1946).
699. — U.S. Patent 2,394,526 (1946); Chem. Abstracts *40,* 5776 (1946).
700. — U.S. Patent 2,423,353 (1947); Chem. Abstracts *41,* 6078 (1947).
701. — U.S. Patent 2,427,314 (1947); Chem. Abstracts *42,* 2286 (1948).

702. — U.S. Patent 2,427,315 (1947); Chem. Abstracts *42,* 1972 (1948).
703. — U.S. Patent 2,461,943 (1949); Chem. Abstracts *43,* 3854 (1949).
704. — U.S. Patent 2,463,471 (1949); Chem. Abstracts *43,* 5051 (1949).
705. — U.S. Patent 2,483,986 (1949); Chem. Abstracts *44,* 3041 (1950).
706. —, and *M. H. Bradley:* U.S. Patent 2,309,681 (1943); Chem. Abstracts *37,* 3769 (1943).
707. —, and *D. W. Kaiser:* U.S. Patent 2,493,703 (1950); Chem. Abstracts *44,* 2574 (1950).
708. — — U.S. Patent 2,535,968 (1950); Chem. Abstracts *45,* 4276 (1951).
709. —, and *D. E. Nagy:* U.S. Patent 2,333,452 (1943); Chem. Abstracts *38,* 2512 (1944).
710. — — U.S. Patent 2,425,287 (1947); Chem. Abstracts *41,* 6769 (1947).
711. — — U.S. Patent 2,427,316 (1947); Chem. Abstracts *42,* 1973 (1948).
712. *Tobiki, H.:* Japan Patent 17,532 (1964); Chem. Abstracts *62,* 5284 (1965).
713. *Tobing, U.,* and *H. Michaud:* German Patent 1,017,363 (1957); Chem. Abstracts *54,* 4059 (1960).
714. *Todd, A. R., F. Bergel, H. L. Fraenkel-Conrat,* and *A. Jacob:* J. Chem. Soc. *1936,* 1601.
715. *Tomhove, W. G.,* and *I. H. Page:* A. M. A. Arch. Internal Med. *106,* 316 (1960).
716. *Tonkin, I. M.:* Brit. J. Pharmacol. *1,* 163 (1946).
717. *Toyoka, H., A. Tutami,* and *K. Ishida:* Nippon Gijutsu Kyokaishi *3,* 79 (1957); Chem. Abstracts *51,* 15161 (1957).
718. *Triggs, W. W.:* Brit. Patent 575,266 (1946); Chem. Abstracts *41,* 5242 (1947).
719. *Tuchmann-Duplessis, H.,* et *L. Mercier-Parot:* Compt. Rend. *253,* 321 (1961).
720. — — J. Ann. Diabetol. *1962,* 141.
721. *Tundermann, W. O.:* U.S. Patent 2,706,179 (1955); Chem. Abstracts *49,* 12862 (1955).
722. *Tybergheim, J. M.,* and *R. H. Williams:* Proc. Soc. Exptl. Biol. Med. *96,* 29 (1957).
723. *Ueda, T.:* Japan Patent 8,237 (1964); Chem. Abstracts *61,* 11900 (1964).
724. — Japan Patent 9034 (1965); Chem. Abstracts *63,* 5613 (1965).
725. —, and *A. Kogya:* J. Soc. Rubber Ind. Japan *29,* 172 (1956); Chem. Abstracts *50,* 17514 (1956).
726. *Ungar, G., L. Freedman,* and *S. L. Shapiro:* Proc. Soc. Exptl. Biol. Med. *95,* 190 (1957).
727. —, *S. Psychayos,* and *H. A. Hall:* Metab., Clin. Exptl. *9,* 36 (1960).
728. *United States Vitamin and Pharmaceutical Corporation:* Brit. Patent 871,237 (1961); Chem. Abstracts *56,* 4676 (1962).
729. *Urbanski, T., B. Skowronska-Serafinowa,* and *H. Dabrowska:* Roczniki Chem. *29,* 450 (1955); Chem. Abstracts *50,* 5548 (1956).
730. — — —, and *J. Jankowska:* Bull. Acad. Polon. Sci. *1,* 74 (1953); Chem. Abstracts *49,* 869 (1955).
731. —, and 17 *Co-Workers:* Gruzlica *26,* 889 (1958); Chem. Abstracts *53,* 4568 (1959).
732. *Voldan, M.:* Cesk. Farm. *1,* 434 (1952); Chem. Abstracts *47,* 5016 (1953).
733. *Volk, B. W.,* and *S. S. Lazarus:* Diabetes *9,* 174 (1960).
734. *Wachter, J. P., R. R. Smeby,* and *A. H. Free:* Am. J. Med. Technol. *26,* 125 (1960).
735. *Wagner, E. C.:* J. Org. Chem. *5,* 133 (1940).
736. *Wagner, W. H.,* u. *H. Vonderbank:* Z. Ges. Exptl. Med. *115,* 66 (1949).
737. *Walker, H. A., S. C. Wilson, C. Farrar,* and *A. P. Richardson:* J. Pharmacol. Exptl. Therap. *104,* 211 (1952).

738. *Walker, R. S.:* Proc. Roy. Soc. Med. *53,* 602 (1960).
739. *Weidenheimer, J. F.,* and *L. Ritter:* U.S. Patent 2,631,146 (1953); Chem. Abstracts *48,* 1443 (1954).
740. *Weinberg, E. D.:* Antibiot. Chemotherapy *11,* 572 (1961).
741. — Antimicrobial Agents Chemotherapy *1961,* 802.
742. —, *R. Chernin,* and *J. H. Billman:* J. Am. Pharm. Assoc., Sci. Ed. *49,* 441 (1960).
742a. *Wellman, K. M., D. L. Harris,* and *P. J. Murphy:* Chem. Commun. *1967,* 568.
743. *Werner, J.:* U.S. Patent 2,390,975 (1945); Chem. Abstracts *40,* 1676 (1946).
744. *Wick, A. N., E. R. Larson,* and *G. S. Serif:* J. Biol. Chem. *233,* 296 (1958).
745. *Williams, R. H., J. S. Hay,* and *M. B. Tjaden:* Ann. N. Y. Acad. Sci. *74,* Art 3, 513 (1959).
746. —, *F. B. Martin, E. D. Henley,* and *H. E. Swanson:* Metab., Clin. Exptl. *8,* 99 (1959).
747. —, *R. H. Pollen, D. C. Tanner,* and *R. H. Barnes:* Ann. Internat. Med. *51,* 1221 (1959).
748. —, and *D. F. Steiner:* Metab., Clin. Exptl. *8,* 548 (1959).
749. —, *D. C. Tanner,* and *W. D. Odel:* Diabetes 7, 87 (1958).
750. —, *J. M. Tybergheim, P. M. Hyde,* and *R. L. Nielsen:* Metab., Clin. Exptl. *6,* 311 (1957).
751. *Williamson, J., D. S. Bertram,* and *E. M. Laurie:* Nature *159,* 885 (1947).
752. *Wilson, C. E.:* U.S. Patent 2,831,810 (1958); Chem. Abstracts *52,* 14151 (1958).
753. *Winnek, P. S.:* U.S. Patent 2,295,884 (1942); Chem. Abstracts *37,* 1229 (1943).
754. *Wittekind, R. W., J. D. Rosenau,* and *G. I. Poos:* J. Org. Chem. *26,* 444 (1961).
755. *Wright, C. I.,* and *J. C. Sabine:* J. Pharmacol. Exptl. Therap. *93,* 230 (1948).
756. *Wright, P.:* Biochem. J. *71,* 633 (1959).
757. *Wystrach, V. P.:* Unpublished results quoted in *J. G. Erikson, P. F. Wiley,* and *V. P. Wystrach,* The Chemistry of Heterocyclic Compounds, Vol. 10, p. 177, *A. Weissberger* (Ed.). New York: Interscience 1956.
758. *Yamada, M., E. Ichikawa,* and *K. Odo:* Yuki Gosei Kagaku Kyokai shi *21,* 946 (1963); Chem. Abstracts *60,* 4146 (1964).
759. *Yavorskii, N. P.:* Farmatsevt. Zh. (Kiev) *19,* 65 (1964); Chem. Abstracts *61,* 9361 (1964).
760. *Yoshikawa, K.,* and *M. Nonaka:* Yushi Kagaku Kyokaishi *2,* 186 (1953); Chem. Abstracts *48,* 9722 (1954).
761. *Young, M. D.:* Trans. Roy. Soc. Med. Hyg. *56,* 252 (1962).
762. *Zabel, M.:* German Patent 1,009,765 (1959); Chem. Abstracts *54,* 13565 (1960).
763. *Zsolnai, T.:* Biochem. Pharmacol. *11,* 995 (1962).

Received November 16, 1967

Library of Congress Catalog Card Number 51-5497. Druck: Meister
Druck, Kassel



Titel-Nr. 4898